# THE
# PERIODIC
# TABLE
## ILLUSTRATED

# THE
# PERIODIC
# TABLE
## ILLUSTRATED

**A Guide to the 118 Chemical Elements**

ABBIE HEADON

**amber**
BOOKS

First published in 2024

Copyright © 2024 Amber Books Ltd

Published by
Amber Books Ltd
United House
London N7 9DP
United Kingdom

www.amberbooks.co.uk
Facebook: amberbooks
YouTube: amberbooksltd
Instagram: amberbooksltd
X(Twitter): @amberbooks

ISBN: 978-1-83886-457-6

Project Editor: Anna Brownbridge
Designers: Rick Fawcett and Mark Batley
Picture Research: Terry Forshaw

Printed in China

# Contents

Introduction...6

Hydrogen...10

Helium...12

Lithium...14

Beryllium...16

Boron...18

Carbon...20

Nitrogen...22

Oxygen...24

Fluorine...26

Neon...28

Sodium...30

Magnesium...32

Aluminum...34

Silicon...36

Phosphorus...38

Sulfur...40

Chlorine...42

Argon...44

Potassium...46

Calcium...48

Scandium...50

Titanium...52

Vanadium...54

Chromium...56

Manganese...58

Iron...60

Cobalt...62

Nickel...64

Copper...66

Zinc...68

Gallium...70

Germanium...72

Arsenic...74

Selenium...76

Bromine...78

Krypton...80

Rubidium...82

Strontium...84

Yttrium...86

Zirconium...88

Niobium...90

Molybdenum...92

Technetium...94

Ruthenium...96

Rhodium...98

Palladium...100

Silver...102

Cadmium...104

Indium...106

Tin...108

Antimony...110

Tellurium...112

Iodine...114

Xenon...116

Caesium...118

Barium...120

Lanthanum...122

Cerium...124

Praseodymium...126

Neodymium...128

Promethium...130

Samarium...132

Europium...134

Gadolinium...136

Terbium...138

Dysprosium...140

Holmium...142

Erbium...144

Thulium...146

Ytterbium...148

Lutetium...150

Hafnium...152

Tantalum...154

Tungsten...156

Rhenium...158

Osmium...160

Iridium...162

Platinum...164

Gold...166

Mercury...168

Thallium...170

Lead...172

Bismuth...174

Polonium...176

Astatine...178

Radon...180

Francium...182

Radium...184

Actinium...186

Thorium...188

Protactinium...190

Uranium...192

Neptunium...194

Plutonium...196

Americium...198

Curium...200

Berkelium...202

Californium...203

Einsteinium...204

Fermium...205

Mendelevium...206

Nobelium...207

Lawrencium...208

Rutherfordium...209

Dubnium...210

Seaborgium...211

Bohrium...212

Hassium...213

Meitnerium...214

Darmstadtium...215

Roentgenium...216

Copernicium...217

Nihonium...218

Flerovium...219

Moscovium...220

Livermorium...221

Tennessine...222

Oganesson...223

Index...224

# Introduction

If you have ever wondered what something is made of – from a flower or a giraffe to the whole Universe itself – the list of ingredients can be found in the periodic table as shown here. This table presents all 118 of the elements that we know to exist, laid out in a way that allows their similarities and differences to be demonstrated visually.

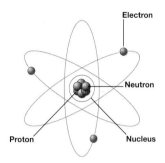

## Names and symbols

Some elements have been known since ancient times, whereas others have only been discovered within the twenty-first century, and as a result, they have a huge variety of names. Element names originate from sources as diverse as astronomical bodies (helium from the Sun, uranium from Uranus), places (europium, californium), people (curium after Marie and Pierre Curie, bohrium after Niels Bohr) and minerals where an element is found (such as samarium from the mineral samarskite).

As well as having a name, each element has a **chemical symbol**, which is an abbreviation of one or two letters. This convenient shorthand enables chemists to describe elements and their compounds concisely. Many of the chemical symbols are

**Atomic numbers**
The smallest basic particle of each element is an atom. Each atom consists of a nucleus, which is made up of protons and neutrons, surrounded by a shell of electrons. Protons have a positive electrical charge, neutrons have no charge at all and electrons have a negative charge. Each atom has the same number of protons and electrons, which means that it has no electrical charge because the positive and negative charges balance each other out. The number of protons in an element's atoms is its atomic number. Hydrogen, the first element in the table, has a single proton in its nucleus, and thus has the atomic number 1. Each subsequent element has an atomic number one step higher than the last, all the way from 1 to 118.

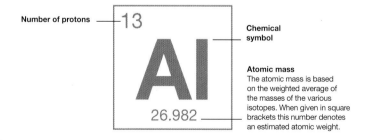

straightforwardly related to their names, such as O for oxygen, H for hydrogen and He for helium. Others are less obvious, being derived from the element's name in Latin or Greek: for example, gold has the symbol Au, from the Latin *aurum*, and mercury is Hg, from the Greek word *hydrargyrum*.

## Isotopes

Although all atoms of any particular element will have the same number of protons in their nuclei, it is possible for elements to have varying numbers of neutrons, and these variations are called isotopes. The most abundant form of carbon on Earth has 6 protons and 6 neutrons in its atomic nucleus, and is known as carbon-12. Two further isotopes of carbon also occur in nature: carbon-13, which has 6 protons and 7 neutrons; and carbon-14, with 6 protons and 8 neutrons. Each of these isotopes has different properties, but they are all carbon and they all share the same position on the periodic table, with atomic number 6.

### The future of the periodic table

At the time of writing, 118 elements have been recognized by the International Union of Pure and Applied Chemistry. Elements 1 to 94 occur naturally on Earth, while elements 95 to 118 have been produced synthetically from nuclear reactions, under laboratory conditions. The discovery of any further elements would lead to an expansion of the table as we currently know it, and there are various competing theories about how these new elements would be arranged. One thing we can be certain of is that scientists are hard at work on this frontier of discovery – and so we can expect to see an updated and expanded periodic table at some point in the future.

Elements are the basic building blocks of the Universe: each one is unique, and none of them can be broken down into anything smaller or simpler. Some are abundant and well known in our daily lives, such as the helium that makes party balloons float, the gold in a wedding ring or the copper piping in domestic heating systems, whereas others are so rare and so radioactive that they have only existed for a few seconds in a laboratory.

**Dmitri Mendeleev**

# The Periodic Table

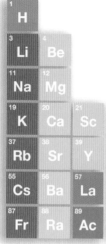

## Origins

The table presented here is a visual representation of the 'periodic law': this means that it organizes the elements in a way that reveals their shared properties. This law was first formulated by the Russian scientist Dmitri Mendeleev in 1869, and his table contained gaps for elements that had not yet been discovered. Their later entry into his table proved that his organizational strategy was correct.

## Groups and Periods

Each vertical column of the table is called a **group**, and these are numbered from 1 on the left of the table to 18 on the right. The elements in each group have the same number of electrons in their outermost shell, and share certain characteristics. For example, the elements in Group 1 each have a single outer electron, and this causes them to be highly reactive with other elements. In contrast, the elements in Group 18 all have the maximum possible number of electrons in their outer shell, and as a result, they are generally inert and unreactive. Because of this characteristic, the elements in Group 18 are known as the 'noble gases'. Each row of the table is called a **period**, and this describes the number of electron shells around the nucleus of the atom of each element. The first shell only holds two electrons, which is why hydrogen and helium have a row to themselves at the top of the table. The second shell holds eight electrons, and so Period 2 contains eight elements. The table can also be divided into four **blocks**, which are defined by the valence electron orbital of each element. The s-block, on the left side of the table, contains hydrogen, helium, the alkali metals and the alkaline-earth metals of Groups 1 and 2. The p-block, on the right side of the table, comprises elements in Groups 13 to 18. The d-block includes the elements from Group 3 through to Group 12. Finally,

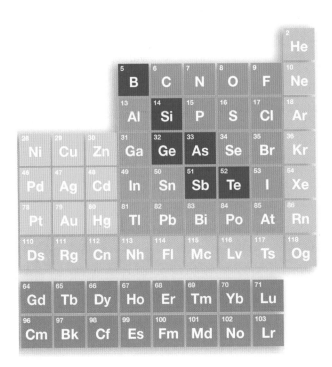

the f-block contains the lanthanide and actinide elements. Although these elements belong between Group 2 and Group 3, they are usually displayed in a double row below the main table in order to save space. Another way of describing the elements is by dividing them into **categories**, which describe shared properties of related elements. These categories have been used as the top-level heading for each element entry in this book.

## Key

Alkali metals

Alkaline earth metals

Transition metals

Post-transition metals

Metalloids

Reactive non-metals

Noble gases

Lanthanides

Actinides

Unknown properties

# Hydrogen

DISCOVERY DATE: 1766    DISCOVERED BY: Henry Cavendish

Hydrogen is the most abundant element in the Universe, making up 75 per cent of the total mass of the Universe. Hydrogen has the simplest structure of all the elements, with a single proton nucleus that is orbited by a single electron. All the other elements in Group 1 of the table are metals, but hydrogen stands apart as a non-metallic gas.

Henry Cavendish, who was the first to identify hydrogen in 1766, also noted in 1781 that it produces water when it is burned. The name hydrogen comes from Greek, and means 'water-former': each water molecule is made up of two atoms of hydrogen bonded to one atom of oxygen, with the chemical formula $H_2O$.

The year 1900 saw the first flight of a rigid, hydrogen-filled airship, an idea promoted by German count Ferdinand von Zeppelin. Such ships became known as 'Zeppelins' and were an extremely popular mode of air transport, with the first non-stop crossing of the Atlantic taking place in 1919. In 1937, the *Hindenberg* airship caught fire over

Protons, which form the nuclei of hydrogen atoms, were created during the first second immediately after the Big Bang, and hydrogen atoms emerged about 370,000 years later.

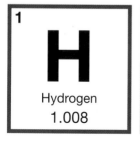

**1**

**H**

Hydrogen

1.008

| | |
|---|---|
| Atomic number: | 1 |
| Group: | Group 1 |
| Period: | Period 1 |
| Block: | s-block |
| Atomic mass: | 1.0080 u |

| | |
|---|---|
| Melting point: | −259.16 °C, −434.49 °F |
| Boiling point: | −252.879 °C, −423.182 °F |
| Density: | 0.000082 g/cm³ (near room temperature) |
| Appearance: | Colourless gas |

New Jersey, with the loss of 36 lives, and this led to the end of commercial travel in hydrogen-filled airships, although later investigations revealed that it was more likely that the fabric coating of the airship had been ignited by static electricity.

The most dangerous application of hydrogen is in the hydrogen bomb (or H-bomb), a more advanced form of the atomic bomb, with the potential to be a thousand times more deadly than the atomic bombs dropped on Hiroshima and Nagasaki in 1945. The hydrogen bomb has never been used in warfare, although a number of nations have carried out tests of such weapons.

The *Hindenberg* airship caught fire and was destroyed while attempting to dock at the Lakehurst Naval Air Station in New Jersey in 1937, with the tragic loss of 36 lives.

# Helium

DISCOVERY DATE: 1895 DISCOVERED BY: Sir William Ramsay (England), Per Teodor Cleve & Nils Abraham Langer (Sweden)

Helium, whose name comes from the Greek word *helios* (meaning 'sun') was formed during the Big Bang, with hydrogen and nitrogen. It is the second most abundant element in the Universe, forming about 24 per cent of the Universe's total elemental mass. Together, hydrogen and helium make up 99 per cent of the observed Universe.

Helium is the first of the noble gases, which form Group 18 of the periodic table. These gases are all characterized by their inertness, or their unwillingness to react with any other element. This comes from the fact that the outer shell of electrons in each atom of these gases is full, and so they have no innate tendency to bond or react with other atoms.

The most important industrial applications of helium are in cryogenics, and as a coolant for superconducting magnets in MRI scanners. On a more domestic level, helium is well known as the gas used for filling party balloons, because it is lighter

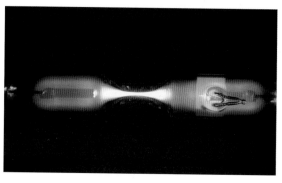

The inert gas helium, seen here in a discharge tube, emits light when electricity is passed through it.

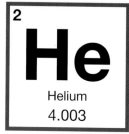

2

## He

Helium

4.003

| | |
|---|---|
| Atomic number: | 2 |
| Group: | Group 18 |
| Period: | Period 1 |
| Block: | s-block |
| Atomic mass: | 4.003 u |

| | |
|---|---|
| Melting point: | Unknown |
| Boiling point: | −268.928 °C, −452.07 °F |
| Density: | 0.000164 g/cm³ (near room temperature) |
| Appearance: | Colourless, odourless gas |

A NASA research team prepare to launch a BARREL balloon – short for Balloon
Array for Radiation belt Relativistic Electron Losses – filled with helium, to explore
the Van Allen belts in space.

than air and therefore enables the balloons to float. It is also used as a lifting gas in
airships, where it has an advantage over another lighter-than-air gas, hydrogen, which
is highly combustible.

If a person inhales helium (perhaps from one of those party balloons), the gas
causes the higher-pitched tones of their voice to resonate more loudly than the lower-
pitched ones, giving a cartoon-character-like effect. However, as helium restricts the
flow of oxygen to the brain, it's best not to try this out too often.

# Lithium

DISCOVERY DATE: 1817     DISCOVERED BY: Johan August Arfwedson

Lithium is the first metal in the periodic table, and along with hydrogen and helium, was created in the Big Bang. It is the lightest of all the metals, and has the lowest density of all solid elements.

As it is a highly reactive and flammable element, lithium has to be stored under a coating of oil or another similar inert liquid, or in a vacuum or inert atmosphere. Its name comes from the Greek word *lithos* (meaning 'stone') to indicate that it was discovered in a mineral, unlike sodium and potassium, which are also alkali metals but that were discovered in living matter.

Lithium has a huge range of applications, despite being corrosive and toxic. In medicine, lithium carbonate is an effective treatment for bipolar disorder, helping to smooth out extremes of mood and manic episodes in patients. It is also used to treat

Lithium, a soft shiny metal, is so reactive that it can ignite spontaneously in air, and it explodes violently on contact with water.

| 3 | |
|---|---|
| **Li** | |
| Lithium | |
| 6.941 | |

| | |
|---|---|
| Atomic number: | 3 |
| Group: | Group 1 |
| Period: | Period 2 |
| Block: | s-block |
| Atomic mass: | 6.941 u |
| Melting point: | 180.50 °C, 356.90 °F |
| Boiling point: | 1330 °C, 2426 °F |
| Density: | 0.534 g/cm³ (near room temperature) |
| Appearance: | Soft silvery-white metal |

severe depression and schizophrenia. However, despite its effectiveness as a psychiatric drug, its toxicity can cause problems such as hypothyroidism and kidney damage, and so its use has to be monitored carefully.

Industrially, the biggest use of lithium is in the production of batteries: in 2020, 65 per cent of all lithium was used for this purpose. Lithium-ion rechargeable batteries are notable for their high energy efficiency and long life, and as electric cars become more widespread, their importance will inevitably increase even further.

An aerial view of lithium fields in the Atacama desert in Chile – a surreal landscape where batteries are born.

# Beryllium

DISCOVERY DATE: 1796    DISCOVERED BY: Louis Nicolas Vauquelin

Beryllium is a soft, light and brittle alkaline-earth metal, and the fourth element in the periodic table. It is far rarer than the three elements that precede it in the table, and can be found in the Earth's crust at a concentration of 2–6 parts per million. Most beryllium is extracted from the mineral beryl, after which is it named. An older alternative name for beryllium was glaucinium, from the Greek word *glykys* (meaning 'sweet') because its compounds had a sweet taste. However, we now know that beryllium is poisonous, and it should not be eaten or inhaled, under any circumstances.

Whereas most metals expand when they get hotter and contract when they get colder, beryllium stays the same at all temperatures. This makes it ideal for use in machinery in which some parts can go through extremes of heat, such as high-speed aircraft.

In nature, beryllium is only ever found in mineral compounds with other elements. This 99.58 per cent pure sample shows elemental beryllium's steel-grey lustre.

| 4 | | |
|---|---|---|
| **Be** | | |
| Beryllium | | |
| 9.012 | | |

| | |
|---|---|
| **Atomic number:** | 4 |
| **Group:** | Group 2 |
| **Period:** | Period 2 |
| **Block:** | s-block |
| **Atomic mass:** | 9.012 u |
| **Melting point:** | 1287 °C, 2349 °F |
| **Boiling point:** | 2469 °C, 4476 °F |
| **Density:** | 1.85 g/cm³ (near room temperature) |
| **Appearance:** | Relatively soft white-grey metal |

An alloy of 2 per cent beryllium and 98 per cent copper creates a high-strength metal used in gyroscopes and other mechanisms. When alloyed with 98 per cent nickel, beryllium forms a metal that does not create sparks, which is ideally suited for use in environments such as oil wells.

Beryllium is relatively transparent to X-rays, due to its low atomic mass and low intensity, and because of this it is the most common material for the windows of X-ray equipment.

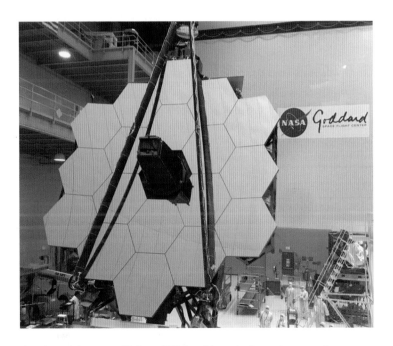

The 18 hexagonal segments of the James Webb Space Telescope's primary mirror are made of beryllium, which is strong, light and maintains its shape at very cold temperatures.

# Boron

DISCOVERY DATE: 1808 DISCOVERED BY: Humphry Davy (UK), Joseph Louis Gay-Lussac & Louis-Jacques Thénard (France)

Boron was discovered independently in 1808 by Sir Humphry Davy in London, and by Joseph Louis Gay-Lussac and Louis-Jacques Thénard in Paris. Its name comes from a combination of the first syllable of borax, which it comes from, and the second syllable of carbon, which it resembles chemically.

In its pure form, boron is a brown amorphous powder. It also exists in a crystalline form as a hard, brittle, shiny metalloid. Amorphous boron is used in fireworks and military flares, because it burns with a vibrant green flame.

Boron is also used in a wide range of compounds for a variety of purposes. Borax, known chemically as sodium borate decahydrate, is used as a bleaching agent in laundry products and as an antiseptic. Boric acid is used in the manufacture of fibreglass and as a flame retardant in insulation, and boric oxide is a key ingredient in the production of the toughened, heat-resistant glass we know as Pyrex.

Boron has several allotropes, including several crystalline forms and the amorphous powder shown in this sample.

| 5 |
|---|
| **B** |
| Boron |
| 10.811 |

| | |
|---|---|
| Atomic number: | 5 |
| Group: | Group 13 |
| Period: | Period 2 |
| Block: | p-block |
| Atomic mass: | 10.811 u |

| | |
|---|---|
| Melting point: | 2076 °C, 3769 °F |
| Boiling point: | 3927 °C, 7101 °F |
| Density: | 2.46 g/cm³ (near room temperature) |
| Appearance: | Brown powder in its amorphous form |

Measuring jug made of Pyrex borosilicate glass, often used in domestic cooking.

Children who play with the toy Silly Putty are also beneficiaries of boron's unique properties. The boric acid in the putty reacts with silicone chains to create a substance that can be squeezed and moulded like a very thick liquid, and that can also bounce like a rubber ball when compressed in the hands.

# Carbon

DISCOVERY DATE: Prehistoric

Carbon is the fifteenth-most abundant on Earth. Its name comes from the Latin word for charcoal, *carbo*, and it has been known since ancient times. It occurs in multiple forms, or allotropes, such as amorphous carbon, graphite, diamond and fullerenes, and all of these forms have different properties.

Carbon is vital for all life on Earth, due to its propensity to create a huge number of compounds – almost 10 million of which are currently known – and to create large macromolecules, which themselves form polymers. The chemistry of carbon is called 'organic chemistry', because many compounds involving carbon are made by or come from living things.

Plants obtain carbon through photosynthesis: they absorb carbon dioxide ($CO_2$) and water ($H_2O$), and use energy from the Sun to split the water into its constituent

Carbon occurs naturally in many different allotropes, depending on how its atoms bond together. These crystalline chunks of graphite are built from thin stacked layers of graphene.

| 6 | |
|---|---|
| **C** | |
| Carbon | |
| 12.011 | |

| | |
|---|---|
| **Atomic number:** | 6 |
| **Group:** | Group 14 |
| **Period:** | Period 2 |
| **Block:** | p-block |
| **Atomic mass:** | 12.011 u |

| | |
|---|---|
| **Melting/boiling point:** | 3642 °C, 6588 °F (sublimes directly from solid to gas) |
| **Density:** | 1.8–2.1 g/cm³ (in its amorphous form), graphite: 2.267 g/cm³ (as graphite), diamond: 3.515 g/cm³ (as diamond) |
| **Appearance:** | Black and opaque (as graphite); transparent (as diamond) |

The burning of fossil fuels in power plants, such as this one, causes vast emissions of carbon dioxide, one of the key factors in climate change.

parts, hydrogen and oxygen. The oxygen is then released back into the atmosphere, and the hydrogen combines with carbon dioxide to form hydrocarbons, which are essential for life. Humans and other animals are unable to photosynthesize, and so we rely on consuming living things in order to obtain the carbon we need.

As well as being the engine house for all life on Earth, carbon plays a fundamental role in providing the fossil fuels methane gas, crude oil and coal. By burning these fuels over the course of human civilization, we have released vast amounts of carbon dioxide into the Earth's atmosphere. Carbon dioxide allows the Sun's light to pass through, but it traps some of the Sun's heat so that it can't escape back into space. At pre-industrial levels, this created a useful 'blanket' around the Earth, keeping us warm in the vastness of the Universe, but we have emitted so much now that a 'greenhouse effect' is causing the planet to become warmer and warmer, with devastating effects on people and the environments in which they live.

# Nitrogen

DISCOVERY DATE: 1772     DISCOVERED BY: Daniel Rutherford

Nitrogen is a colourless and odourless gas that makes up about 78 per cent of the Earth's atmosphere, at about 4000 trillion tons. It usually occurs as a two-atom molecule, N2. Although it was first isolated during the 1760s, independently by Henry Cavendish and Joseph Priestley (who both removed oxygen from air and noted that the remaining gas would extinguish a flame and cause small animals to die of asphyxiation), it was not until 1772 that another scientist, Daniel Rutherford, identified it as an element.

The English word 'nitrogen' was derived from the French *nitrogène*, combining 'nitre' (also known as potassium nitrate or saltpetre) and the suffix *-gène* (meaning 'forming'). This is because nitrogen is an essential ingredient of nitric acid, which comes from nitre. The German word for nitrogen, *Stickstoff*, is a more direct reference

The mineral nitratine is a naturally occurring form of sodium nitrate, $NaNO_3$.

| 7 | |
|---|---|
| **N** | |
| Nitrogen | |
| 14.007 | |

| | |
|---|---|
| **Atomic number:** | 7 |
| **Group:** | Group 15 |
| **Period:** | Period 2 |
| **Block:** | p-block |
| **Atomic mass:** | 14.007 u |
| **Melting point:** | −209.86 °C, −345.75 °F |
| **Boiling point:** | −195.795 °C, −320.431 °F |
| **Density:** | 0.001251 g/cm³ (near room temperature) |
| **Appearance:** | Colourless gas, liquid or solid |

to its propensity to choke (*ersticken*) living creatures. Group 15 of the table, which has nitrogen in its uppermost row, is called the pnictogens, from a Greek root word meaning to choke or strangle.

Nitrogen is essential for all living things, as it occurs in DNA, RNA, amino acids and in adenosine triphosphate, a molecule that provides energy for processes in living cells, such as muscle contraction and chemical synthesis. This role is just one stage of the natural nitrogen cycle: when plants and animals die, their nitrogen returns to the soil, ready to be taken up again by plants and consumed by animals.

Nitrogen is also vital commercially, for use in fertilizers, as dry ice, as an inert gas for food storage, in compounds such as Kevlar and in the manufacture of stainless steel.

**Racing cars in the National Hot Rod Association's Top Fuel class are powered by nitromethane fuel, and can reach speeds of up to 150 metres per second.**

# Oxygen

DISCOVERY DATE: 1774     DISCOVERED BY: Joseph Priestley (England) & Carl Wilhelm Scheele (Sweden)

Oxygen – element number eight in the periodic table – is a highly reactive non-metal that easily forms oxides with most elements. Despite its tendency to react with other elements, it is found pure in the natural world, usually as a molecule of two oxygen atoms (chemical formula $O_2$). Oxygen is the third most abundant element in the Universe after hydrogen and helium, and currently makes up 20.95 per cent of the Earth's atmosphere. This level has varied during our planet's history; in fact, for about the first two billion years of the Earth's existence, there was no free oxygen in the atmosphere. It was only after the evolution of photosynthesizing organisms around two and half billion years ago that oxygen was emitted into the air. This led to the extinction of many earlier life forms, and laid the pathway for life as we know it today.

The oxygen concentration in the Earth's atmosphere hit a peak of 35 per cent some 300 million years ago, during the Carboniferous period. This abundance of oxygen is

When oxygen molecules enter an electronically excited state, they change colour. This phenomenon can be observed not only in a laboratory but also in the night sky, in the form of the Northern and Southern Lights.

| | |
|---|---|
| **8** | |
| **O** | |
| Oxygen | |
| 15.999 | |

| | |
|---|---|
| **Atomic number:** | 8 |
| **Group:** | Group 16 |
| **Period:** | Period 2 |
| **Block:** | p-block |
| **Atomic mass:** | 15.999 u |

| | |
|---|---|
| **Melting point:** | −218.79 °C, −361.82 °F |
| **Boiling point:** | −182.962 °C, −297.332 °F |
| **Density:** | 0.001429 g/cm³ (near room temperature) |
| **Appearance:** | Colourless gas; pale blue liquid or solid |

Scuba divers rely on tanks containing a mixture of oxygen and nitrogen in order to be able to breathe comfortably underwater.

thought to be a factor in the existence of giant insects, far bigger than ones we know today. The fossil record shows that dragonfly-like creatures with a wingspan of 75cm (30in) were prevalent during this period.

Oxygen is required by all plants, animals and fungi for the extraction of energy from food in a process called cellular respiration, which processes carbon dioxide as its waste product. Plants then absorb the carbon dioxide, and through photosynthesis involving sunlight and water, produce carbohydrates and release oxygen back into the air. Humans rely on a consistent supply of oxygen to stay alive, and if the oxygen concentration of the air around us falls to 17 per cent or less, we are unlikely to survive.

# Fluorine

DISCOVERY DATE: 1886     DISCOVERED BY: Henri Moissan

Fluorine's name comes from the Latin word *fluo*, which means 'to flow'. Although fluorine in its pure form is a gas at room temperature, and a solid when found occurring naturally in the mineral fluorite, it gains its flowing name from the fact that, from the sixteenth century onwards, it was added to metal ores to reduce their melting points during smelting.

Fluorine is a dangerously toxic gas, and the most reactive of all the elements: the only elements it does not react with are the noble gases in Group 18 of the periodic table. Its high reactivity made it very difficult for the first scientists who tried to isolate it from the more stable minerals that contain it, and in fact, some of them even died as a result of their experiments. The work of these 'fluorine martyrs' paved the way for a French chemist, Henri Moissan, to finally achieve a safe isolation of the element via electrolysis in 1886, a feat for which he was awarded the Nobel Prize in Chemistry in 1906.

The mineral fluorite was first discovered at Rogerley Mine in County Durham, England, in around 1970 in the UK. Rare earth elements within the mineral cause the green crystals to glow purple in daylight.

| 9 | | |
|---|---|---|
| **F** | | |
| Fluorine | | |
| 18.998 | | |

| | |
|---|---|
| **Atomic number:** | 9 |
| **Group:** | Group 17 |
| **Period:** | Period 2 |
| **Block:** | p-block |
| **Atomic mass:** | 18.998 u |

| | |
|---|---|
| **Melting point:** | −219.67 °C, −363.41 °F |
| **Boiling point:** | −188.11 °C, −306.60 °F |
| **Density:** | 0.001696 g/cm³ (near room temperature) |
| **Appearance:** | Pale yellow gas; bright yellow liquid |

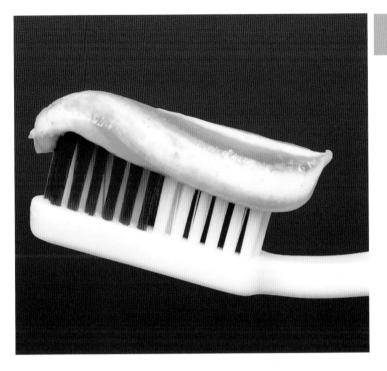

Various forms of fluoride, including sodium fluoride (NaF) and stannous fluoride ($SnF_2$), have been shown to prevent cavities and dental erosion when added to toothpaste.

Despite its dangerous qualities, fluorine is widely used in industry and plays more of a role in our daily lives than we might imagine. The non-stick coating in our frying pans, best known by the trade name Teflon, is actually a fluoropolymer called polytetrafluoroethylene (PTFE), and other fluorine-based polymers provide us with water-resistant rainwear, electrical insulation and films for solar cells.

Fluoride, an inorganic, monatomic anion of fluorine, plays a vital role in dental health by preventing tooth decay and reducing cavities. Approximately 6 per cent of the world's population receives fluoride directly through their water supply, and fluoride is a ubiquitous constituent of toothpastes and mouthwashes. Although some people are opposed to fluoridation of water supplies, evidence to date suggests that it reduces tooth decay and does not cause any significant health problems.

# Neon

DISCOVERY DATE: 1898     DISCOVERED BY: Sir William Ramsay and Morris Travers

Neon is the fifth-most abundant element in the Universe (after hydrogen, helium, oxygen and carbon), but on Earth it is only found in trace amounts. It was first discovered by Sir William Ramsay and Morris Travers, along with the gases krypton and xenon, by the fractional distillation of liquid air, and this same method is used to produce neon for commercial uses today, due to its rarity in our atmosphere.

William Ramsay's son suggested 'novum' as the name for this element, from the Latin word *novus* (meaning 'new') but his father preferred to use the Greek word *neos*, with the same meaning, as the base for the element we now know as neon.

Neon gas glows a bright red-orange colour when electricity is passed through it, and neon lights have dominated street advertising ever since Georges Claude showed off the first illuminated neon tube in 1910 at the Paris Motor Show. A rainbow of other coloured lights can be created by the addition of other gases, such as mercury,

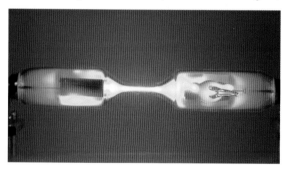

Neon gas glows with an orange-red colour in this discharge tube.

**10**

# Ne

Neon

20.180

| | |
|---|---|
| Atomic number: | 10 |
| Group: | Group 18 |
| Period: | Period 2 |
| Block: | p-block |
| Atomic mass: | 20.180 u |

| | |
|---|---|
| Melting point: | −248.59 °C, −415.46 °F |
| Boiling point: | −246.046 °C, −410.883 °F |
| Density: | 0.0009 g/cm³ (at standard temperature and pressure) |
| Appearance: | Colourless, odourless gas |

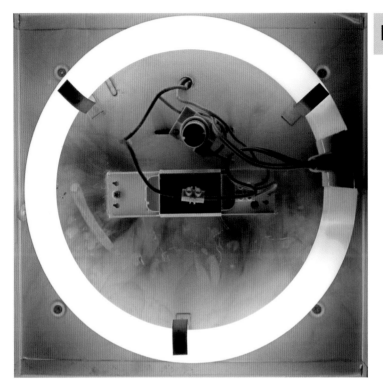

Tubes filled with neon gas provide lighting in a huge range of situations, from advertising signs to ceiling lamps.

carbon dioxide and helium. As well as being famously used for lighting, neon is also used in high-voltage switching gear, lasers, vacuum tubes and diving gear. Another important use of neon is in low-temperature refrigeration, as it can efficiently maintain a temperature of −246°C (-410.80°F).

# Sodium

DISCOVERY DATE: 1807    DISCOVERED BY: Humphry Davy

Sodium is a member of Group 1 of the periodic table, and possesses the reactive qualities of all the metals in this group. In its pure form it is a silvery-white soft metal that tarnishes rapidly when exposed to the air and it also reacts vigorously when in contact with water. Its chemical symbol is Na, from the Latin word *natrium* (meaning 'soda').

Sodium is the sixth most abundant element in the Earth's crust, and its most common compound is sodium chloride (NaCl), also familiar to us as table salt. Salt's flavour and value as a food preservative have been known for thousands of years, and the word 'salary' has its origins in the *salarium argentum* or 'salt money' that was sometimes paid to Roman soldiers. Salt also appears in many other idioms, such as 'salt of the earth', 'to be worth one's salt' and 'take something with a pinch of salt' – linguistically, we can't seem to get enough of this tasty and significant white stuff.

Halite (or rock salt) is the natural mineral form of sodium chloride (NaCl), and takes the form of isometric crystals.

| 11 | | |
|---|---|---|
| **Na** | | |
| Sodium | | |
| 22.990 | | |

| | |
|---|---|
| Atomic number: | 11 |
| Group: | Group 1 |
| Period: | Period 3 |
| Block: | s-block |
| Atomic mass: | 22.990 u |
| Melting point: | 97.794 °C, 208.029 °F |
| Boiling point: | 882.940 °C, 1621.292 °F |
| Density: | 0.968 g/cm³ (near room temperature) |
| Appearance: | Soft silvery metal |

Salt is an essential ingredient in the human diet. It helps to regulates the fluid balance of our bodies, and it transmits electrical signals in our nervous systems. However, over-consumption of sodium is associated with increased risk of many health problems, including stroke, cardiovascular disease, high blood pressure and kidney disease. According to the World Health Organization, we should aim to consume no more than 5g (0.17oz) of salt per day, but the global mean intake of salt is currently 10.78g (0.38oz) per day, more than double the recommended maximum.

**Salt has countless uses, including as a flavour enhancer, tenderizer and food preservative. Without salt, our meals just wouldn't taste the same.**

# Magnesium

DISCOVERY DATE: 1755     DISCOVERED BY: Joseph Black

A silvery-white metal and member of the alkaline-earth metals (Group 2 of the periodic table), magnesium burns easily in air, with a bright white light. Because of this property, it is used in fireworks and flares, and it also plays a role on stage to create theatrical effects such as lightning and ghostly appearances.

Magnesium is present in the chlorophyll molecule, which gives green plants their colour and helps them to absorb energy from the Sun during the process of photosynthesis. As it is so widespread in the plant world, magnesium is also present in nearly all living animals, absorbed either by eating plants or the animals that ate them.

In the human body, magnesium plays a key role as a cofactor in more than 300 enzyme systems. It also contributes to the development of human bones and assists with the synthesis of DNA and RNA. Good sources of dietary magnesium are vegetables, nuts, cereals, cocoa and spices.

Magnesium is a shiny silver metal two-thirds as dense as aluminium. It has the lowest melting point and boiling point of all the alkaline-earth metals.

## 12
## Mg
Magnesium
24.305

| Atomic number: | 12 |
| --- | --- |
| Group: | Group 2 |
| Period: | Period 3 |
| Block: | s-block |
| Atomic mass: | 24.305 u |

| Melting point: | 650 °C, 1202 °F |
| --- | --- |
| Boiling point: | 1091 °C, 1994 °F |
| Density: | 1.738 g/cm$^3$ (near room temperature) |
| Appearance: | Silvery-white metal |

Magnesium alloy wheels are manufactured by casting or forging. Forged wheels are usually lighter and stronger than cast wheels, but more expensive.

After iron and aluminium, magnesium is the third most commonly used structural metal. Its lightness and strength makes it ideal for use in alloys for automotive and aircraft engineering, and its electrical properties mean that it is used in mobile phones, laptops, cameras and other electronic devices where lightness is a desirable quality.

# Aluminium

DISCOVERY DATE: 1825     DISCOVERED BY: Hans Oersted

Aluminium is the most abundant metal in the Earth's crust, at 8.1 per cent, and the third most abundant element, after oxygen and silicon. It is rarely found in a pure form, due to its propensity to form oxides and silicates. This explains why there is more aluminium, proportionately, in the Earth's crust than in the Universe as a whole, because less reactive elements have already sunk into the Earth's core.

Aluminium's name comes from *alumen*, the Latin word for alum, a mineral that contains aluminium. Lively debates took place about the exact name that should be given to this element, with Jöns Jacob Berzelius proposing the name aluminium in 1811 and Humphry Davy using the spelling aluminum in a textbook in 1812. To this day, the difference remains, with the USA and Canada preferring the shorter aluminum to the rest of the world's aluminium.

**Pure aluminium is so shiny that it rivals silver as a metal mirror of visible light. When exposed to air, it forms a protective layer of oxide on its surface.**

**13**

# Al

Aluminium

26.982

| | |
|---|---|
| **Atomic number:** | 13 |
| **Group:** | Group 13 |
| **Period:** | Period 3 |
| **Block:** | p-block |
| **Atomic mass:** | 26.982 u |

| | |
|---|---|
| **Melting point:** | 660.32 °C, 1220.58 °F |
| **Boiling point:** | 2470 °C, 4478 °F |
| **Density:** | 2.70 g/cm³ (near room temperature) |
| **Appearance:** | Silvery-grey metal |

Although it has no biological role in animal nutrition, aluminium is absorbed by all plants and is therefore consumed by most of us every day in small amounts. Nearly all of it passes straight through our digestive systems, and it is only problematic if it enters the bloodstream, where it may be associated with an increased risk of Alzheimer's disease, although this link has not been proven.

Aluminium is an ideal metal for manufacturing purposes because it is light, corrosion-resistant and easy to shape. It is usually used in an alloy with a small percentage of another metal in order to provide the strength it lacks, and the most common of these alloying metals are copper, zinc, magnesium, manganese and silicon. Its uses range from aircraft fuselages and window frames to drinks cans, cooking utensils and kitchen foil.

Conceived by Douglas' leading aeronautical designer Arthur Raymond, the DC-3 incorporated numerous aeronautical advancements. These included a strong, yet relatively light, highly streamlined, all-aluminum, semi-monocoque fuselage.

# Silicon

DISCOVERY DATE: 1824    DISCOVERED BY: Jöns Jacob Berzelius

Silicon is the eighth most abundant element in the Universe, and the second most abundant in the Earth's crust after oxygen. It does not occur in its pure form in the natural world, instead being found in silicon oxide (also known as silica) and silicates. Its name comes from the Latin word for 'flint' (*silex* or *silicis*) as flint is made of silica.

For a great many of its commercial uses, silicon does not need to be separated from its mineral compounds. Silicate minerals include clay, silica sand and most types of building stone. Silicates are also vital in the production of cement and concrete – so it's fair to say that without silicon, our architectural surroundings would be very different. We also depend on silica to provide us with glass for our windows, drinking glasses and spectacles.

Pure silicon is a blue-grey shiny metallic semiconductor. As with typical semiconductors, when its temperature rises, its resistivity decreases.

| 14 | |
|---|---|
| **Si** | |
| Silicon | |
| 28.085 | |

| Atomic number: | 14 |
|---|---|
| **Group:** | Group 14 |
| **Period:** | Period 3 |
| **Block:** | p-block |
| **Atomic mass:** | 28.085 u |

| **Melting point:** | 1414 °C, 2577 °F |
|---|---|
| **Boiling point:** | 3265 °C, 5909 °F |
| **Density:** | 2.3290 g/cm³ (near room temperature) |
| **Appearance:** | Crystalline with a blue-grey sheen |

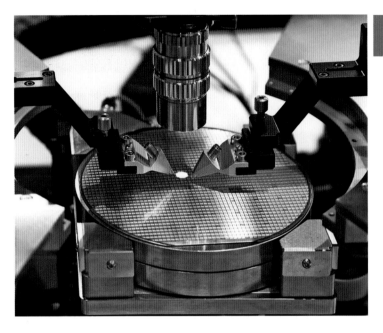

**Integrated circuit boards are constructed on a supporting wafer of silicon, as shown here in this device, which controls the manufacture of microcircuits.**

As well as fulfilling our needs at this relatively large scale, silicon fulfils another completely different role at the heart of our technological devices. As an electrical semiconductor, pure silicon is essential to the functioning of computers, smartphones and other devices involving microelectronics. We often refer to the past few decades as the Information Age or Digital Age, but another equally appropriate name for the period we are living in is the Silicon Age.

Silicon is also used in the production of synthetic polymers called silicones. These usually take the form of colourless oils or rubbery substances, and they are useful as sealants, adhesives and in electrical and thermal insulation. In the medical field, silicone is used in implants, contact lenses and scar treatment sheets.

# Phosphorus

DISCOVERY DATE: 1669      DISCOVERED BY: Hennig Brand

Anyone who associates the idea of science with white lab coats, high-tech equipment and a generally sterile environment is likely to have their illusions crushed by the story of the discovery of phosphorus. In 1669, a German alchemist named Hennig Brand was trying to create the 'philosopher's stone', a mythical substance that could turn common base metals into silver or gold, and his method was to boil large quantities of human urine, to see what happened. What in fact happened was that he created a waxy white substance that glowed with an eerie light in darkness. This substance was white phosphorus, and its light-giving quality gave it its name, as the Greek word *phosphoros* means 'light-bringer'. Brand thought that the element he had found might be useful as a source of light, but it caught fire so easily that it was much too dangerous for this purpose.

Apatite, or inorganic phosphate rock, is the chief commercial source of phosphorus today. Major reserves are found in Morocco, Algeria and Tunisia.

| 15 P Phosphorus 30.974 | | |
|---|---|---|
| **Atomic number:** | 15 | |
| **Group:** | Group 15 | |
| **Period:** | Period 3 | |
| **Block:** | p-block | |
| **Atomic mass:** | 30.974 u | |
| **Melting point:** | 44.15 °C, 111.5 °F (white phosphorus) ~590 °C, ~1090 °F (red phosphorus) | |
| **Boiling point:** | 280.5 °C, 536.9 °F (white phosphorus) | |
| **Density:** | 1.823 g/cm³ (white phosphorus, near room temperature), ≈2.2–2.34 g/cm³ (red phosphorus, near room temperature) | |
| **Appearance:** | Waxy solid (white phosphorus), amorphous solid (red phosphorus) | |

In the nineteenth century, the flammability of phosphorus led to its use as a coating for the tips of matches, but its toxicity caused terrible injuries to the factory workers who were forced to breathe phosphorus fumes for 14-hour shifts. In 1888, the female factory workers at Bryant & May went on strike, bringing production to a halt and the company was forced to improve safety standards. White phosphorus was eventually banned in match production, and today's matchboxes have a striking surface made up of 50 per cent red phosphorus instead.

Phosphorus has been used as a weapon of war, most horrifyingly in the Allied air raids of July 1943 on Hamburg – the city where Hennig Brand first made his discovery. However, the combination of this deadly element with oxygen creates phosphates ($PO_4^{3-}$) that are vital for all known forms of life.

Red phosphorus on the striking surface of a matchbox has replaced the toxic white phosphorus once used on the tips of matches.

# Sulfur

Sulfur's name can be traced back to two separate roots, the Latin *sulfurium* or the Sanskrit *sulvere*, however, it has a host of other names, too, due to the fact that humans have known about this element since antiquity. In English, sulfur is also known as brimstone, meaning 'burning stone'. It should be mentioned that users of British English often prefer the spelling 'sulphur', although the International Union of Pure and Applied Chemistry recommends the 'f' spelling for global use.

Sulfur can be found in its pure form on Earth, without needing further refinement, but for industrial purposes today, most sulfur is obtained from oil, natural gas and tar sands, in the form of hydrogen sulfide. Approximately 85 per cent of the global production of sulfur goes into sulfuric acid, which is vital for the extraction of phosphate ores used in the manufacture of fertilizers. Sulfur is also used in the production of matches, and as an insecticide and fungicide.

Sulfur crystals have a characteristic yellow colour, but when burned, they emit a blood-red liquid and a blue flame.

| 16 | |
|---|---|
| **S** | |
| Sulphur/Sulfur | |
| 32.06 | |

| | |
|---|---|
| **Atomic number:** | 16 |
| **Group:** | Group 16 |
| **Period:** | Period 3 |
| **Block:** | p-block |
| **Atomic mass:** | 32.06 u |

| | |
|---|---|
| **Melting point:** | 115.21 °C, 239.38 °F |
| **Boiling point:** | 444.6 °C, 832.3 °F |
| **Density:** | 1.960 g/cm³ (near room temperature) |
| **Appearance:** | Yellow crystals or powder |

The African American writer Booker T. Washington described a sulfur mine in Sicily as 'about the nearest thing to hell that I expect to see in this life'. This mine at Kawah Ijen in Indonesia paints a similar picture.

Sulfur dioxide and sulfites are used widely as food preservatives, as they can kill bacteria while also preventing food from oxidizing and turning brown.

The human body contains, on average, 140g (4.9oz) of sulfur, where it plays a vital role as part of the essential amino acid methionine. Good sources of dietary sulfur include meat, fish, eggs, nuts, chickpeas and members of the allium family of vegetables (such as onions, leeks, shallots and garlic). While being essential for our survival, sulfur can also play a more antisocial role as the key ingredient in bad breath, when bacteria in the mouth emit volatile sulfur compounds. Maintaining good dental hygiene is the best way to keep this sulfurous odour at bay.

# Chlorine

The name 'chlorine' comes from the Greek word *chloros*, meaning 'greenish yellow', which refers to the colour of this element. It was first discovered in 1774 by the Swedish chemist Carl Wilhelm Scheele, although it was not until 1810 that the British scientist Humphry Davy proved that it was an element.

Chlorine is never found in its pure form in nature, as it can combine directly with almost every other element. It is commercially produced through the electrolysis of brine (a solution of NaCl, otherwise known as table salt), where the chloralkali process leads to the production of chlorine gas, hydrogen gas and sodium hydroxide.

Chlorine is highly corrosive and can be deadly for almost all forms of life, and this gives it a paradoxical role as an element that is both highly dangerous and incredibly

Halite (or rock salt) is made from a compound of sodium (Na) and chlorine (Cl). It is found in sedimentary rocks formed from the evaporation of seawater.

| 17 Cl | | |
|---|---|---|
| Chlorine | | |
| 35.45 | | |

| | |
|---|---|
| Atomic number: | 17 |
| Group: | Group 17 |
| Period: | Period 3 |
| Block: | p-block |
| Atomic mass: | 35.45 u |

| | |
|---|---|
| Melting point: | −101.5 °C, −150.7 °F |
| Boiling point: | −34.04 °C, −29.27 °F |
| Density: | 0.0032 g/cm³ (near room temperature) |
| Appearance: | Pale yellow-green gas |

useful. Even a small amount in the air irritates the eyes and throat, and it was weaponized to horrific effect during World War I (1914–18) as a poison gas. However, this very toxicity gives chlorine the power to kill bacteria, making it an effective disinfectant. Chlorine is used to make bleach, to keep swimming pool water clean and to purify drinking water. In this last role, through killing many waterborne diseases, chlorine has saved far more lives than it has taken as a weapon.

Polyvinyl chloride, better known to us as PVC, is just one of the wide range of products containing chlorine that we use in our daily lives, such as here in these PVC drainpipes.

# Argon

DISCOVERY DATE: 1894          DISCOVERED BY: Lord Rayleigh and William Ramsay

Argon is the third most abundant gas in the Earth's atmosphere at 0.93 per cent, after nitrogen (78.08 per cent) and hydrogen (20.95 per cent). Its name comes from the Greek word *argos* (meaning 'lazy') due to the fact that it does not combine with other elements. (The PR department for argon was obviously taking a day off when this name was made official, unlike Group 18 as a whole, which carries the rather more positive-sounding title of 'the noble gases' to describe this very same unwillingness to react.)

This inertness makes argon useful to us in many different ways. It is used to fill the spaces between the panes of double-glazed windows, where it helps to prevent heat passing through. It can replace normal air in museum display cases to prevent documents from being damaged by mould and germs. One example of this is at the Library of Congress in Washington, DC, USA, where an argon-filled case protects

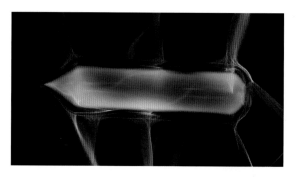

When excited in a discharge tube, argon gas gives off a purple-blue glow.

| 18 | |
|---|---|
| **Ar** | |
| Argon | |
| 39.95 | |

| | |
|---|---|
| Atomic number: | 18 |
| Group: | Group 18 |
| Period: | Period 3 |
| Block: | p-block |
| Atomic mass: | 39.95 u |

| | |
|---|---|
| Melting point: | −189.34 °C, −308.81 °F |
| Boiling point: | −185.848 °C, −302.526 °F |
| Density: | 0.001784 g/cm³ (near room temperature) |
| Appearance: | Colourless gas |

the 1507 world map compiled by German cartographer Martin Waldseemüller, the first map to ever use the word 'America', which is sometimes called 'America's birth certificate'. It is also important as a coolant and a fire extinguisher in places where water or foam would damage sensitive equipment.

Just like neon, which sits immediately above it in the periodic table, argon can be used in lighting. Pure argon produces a violet or lilac-coloured light, whereas a mixture of argon and mercury gives a blue light. Argon is also often used with mercury as a filling for low-energy lightbulbs.

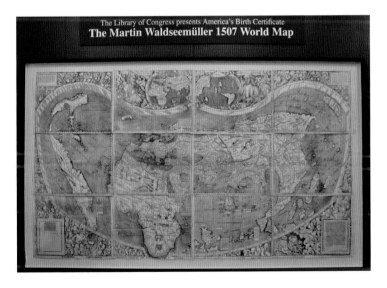

Argon is used in museum display cases to protect old valuable documents, such as the Martin Waldseemüller 1507 World Map in Washington, DC, USA.

# Potassium

DISCOVERY DATE: 1807     DISCOVERED BY: Humphry Davy

Potassium is a soft, silver-coloured metal that tarnishes rapidly on exposure to air. In order to prevent this, it is often stored under a protective layer of oil or grease. Just like all the other members of Group 1 of the periodic table, potassium is highly reactive, and it will catch fire if it comes into contact with water, burning with a lilac-coloured flame. The element's name comes from the English word 'potash', a mixture of plant or wood ashes soaked in water, which was used as far back as the Bronze Age for bleaching fabrics, making glass and ceramics and producing soap. Its symbol, K, is derived from the Latin word *kalium*, which itself comes from the Arabic word *qali*, meaning 'alkali'.

Potassium is needed by almost all living things, and in the human body it carries out many functions, such as maintaining the balance of fluids and electrolytes, and helping nerves and muscles to work properly. It also plays a role in maintaining

Potassium salts such as this red mineral, which also contains layers of sodium salt, are used in the commercial production of potassium.

| 19 K Potassium 39.098 | | |
|---|---|---|
| **Atomic number:** | 19 | |
| **Group:** | Group 1 | |
| **Period:** | Period 4 | |
| **Block:** | s-block | |
| **Atomic mass:** | 39.098 u | |
| **Melting point:** | 63.5 °C, 146.3 °F | |
| **Boiling point:** | 757.643 °C, 1395.757 °F | |
| **Density:** | 0.89 g/cm³ (near room temperature) | |
| **Appearance:** | Soft silvery metal | |

healthy bones, as the potassium salts found in fruit and vegetables help to retain calcium and acid in the bones, which is especially important in the prevention of osteoporosis. Some of the foods most abundant in potassium are potatoes, yams, nuts, bananas, avocados, dried apricots and – good news for anyone with a sweet tooth – chocolate.

Bananas, nuts and dried fruits are all good sources of essential dietary potassium.

# Calcium

DISCOVERY DATE: 1808      DISCOVERED BY: Humphry Davy

Calcium is the fifth most abundant element in the Earth's crust, at 4.1 per cent, and the third most abundant metal, after iron and aluminium. Due to its reactivity, it is not found in its pure form in the natural world, but in compounds such as calcium carbonate, calcium sulfate and calcium fluoride, which are known more commonly as limestone, gypsum and fluorite respectively. The name 'calcium' comes from the Latin word *calx*, meaning 'lime', which is a product derived from heating limestone.

Calcium is vital for almost all living things, particularly for the growth of healthy bones and teeth. The average human body contains 1–1.2kg (2.2–2.6lb) of it, and nearly all of this is stored in our bones. The rest is used to support the synthesis and

Calcite, a naturally occurring form of calcium carbonate ($CaCO_3$), can be found in more than 1000 different crystallographic forms.

| 20 | |
|----|----|
| **Ca** | |
| Calcium | |
| 40.078 | |

| | |
|---|---|
| **Atomic number:** | 20 |
| **Group:** | Group 2 |
| **Period:** | Period 4 |
| **Block:** | s-block |
| **Atomic mass:** | 40.078 u |

| | |
|---|---|
| **Melting point:** | 842 °C, 1548 °F |
| **Boiling point:** | 1484 °C, 2703 °F |
| **Density:** | 1.55 g/cm³ (near room temperature) |
| **Appearance:** | Soft silvery-white metal |

At Pamukkale in south-western Turkey, calcium deposits from hot springs have formed these dramatic terraced pools made from travertine, a sedimentary rock.

function of blood cells, regulating muscle contractions, blood clotting and nerve conduction. Good sources of dietary calcium include cheese, milk, green leafy vegetables, bread and the kinds of fish whose bones can be eaten, such as sardines and whitebait.

Pure calcium is used in the manufacture of steel, and in alloys with aluminium, beryllium, copper, lead and magnesium. Limestone is used as a building material, continuing a tradition that stretches back 4500 years to the ancient Egyptian pyramids, which were constructed from this material, and many of its derivatives such as quicklime and slaked lime are used to make cement and plaster. So overall, it's fair to say that without calcium, both our homes and our bodies would be very structurally unsound.

# Scandium

DISCOVERY DATE: 1879      DISCOVERED BY: Lars Fredrik Nilson

When he first created his periodic table of the elements in 1869, Dmitri Mendeleev left a space for this element, being convinced that a new lightweight metal would be identified at a later date – and he was proved right 10 years later. In 1879, Swedish scientist Lars Fredrik Nilson isolated a tiny amount of scandium oxide, and he named the newly discovered element scandium after his native Scandinavia. It was not until 1937 that pure scandium was obtained for the first time through the electrolysis of scandium chloride.

Globally, 15 to 20 tonnes (16.5 to 22 tons) of scandium are produced each year, with both demand and production increasing. It is almost as light as aluminium but has a higher melting point, and this makes it ideal for use in aircraft manufacture. Aluminium-scandium alloys were used in the construction of the Russian MiG-21

This sample of 99.99 per cent pure scandium has a silvery surface. This would oxidize with a yellow or pink-ish surface if exposed to the air.

| 21 |
|----|
| **Sc** |
| Scandium |
| 44.956 |

| | |
|---|---|
| **Atomic number:** | 21 |
| **Group:** | Group 3 |
| **Period:** | Period 4 |
| **Block:** | d-block |
| **Atomic mass:** | 44.956 u |
| **Melting point:** | 1541 °C, 2806 °F |
| **Boiling point:** | 2836 °C, 5136 °F |
| **Density:** | 2.985 g/cm³ (near room temperature) |
| **Appearance:** | Silvery-white metal |

This MiG-29 Polish Air Force fighter jet is partly made from aluminium-scandium alloys.

and MiG-29 military aircraft, and these alloys are also highly valued in the production of sports equipment such as lacrosse sticks, baseball bats and bicycle frames.

Scandium-46, a radioactive isotope of the element, was used by the Stasi – the East German secret police – during the 1970s and 1980s to track suspected dissidents. Agents would spray a solution of scandium-46 onto the floors of rooms where they thought suspects would meet, or onto bank notes, and then they would wear Geiger counters strapped to their bodies and try to track people's movements. This was the first well-documented case of such a plan being put into practice, and it is certain that it endangered the health of everyone involved.

# Titanium

DISCOVERY DATE: 1791    DISCOVERED BY: William Gregor

Titanium, the ninth most abundant element in the Earth's crust, takes its name from the Titans, the generation of Greek gods who preceded Zeus, Hades and the other Olympian gods. The Titans were fabled for their strength, just like the metal named after them.

Titanium is as strong as steel while being much lighter and less dense. When alloyed with metals such as aluminium, molybdenum and iron, it is a valuable material for the construction of aeroplanes, missiles and spacecraft, where lightness, strength and resistance to corrosion and heat are vital qualities.

Titanium alloys are also used in the production of golf clubs and other types of sports equipment, bicycle frames and crutches – and its flexibility means that it can be made into glasses frames that spring back into shape if you accidentally sit on them.

Ilmenite is a common titanium-containing ore and was the mineral in which William Gregor first identified titanium as a new element.

| 22 | |
|---|---|
| **Ti** | |
| Titanium | |
| 47.867 | |

| | |
|---|---|
| **Atomic number:** | 22 |
| **Group:** | Group 4 |
| **Period:** | Period 4 |
| **Block:** | d-block |
| **Atomic mass:** | 47.867 u |

| | |
|---|---|
| **Melting point:** | 1668 °C, 3034 °F |
| **Boiling point:** | 3287 °C, 5949 °F |
| **Density:** | 4.506 g/cm³ (near room temperature) |
| **Appearance:** | Hard silvery metal |

Because titanium connects well with bone and is not rejected by the human body, it is also used successfully in artificial joint replacements and dental implants.

One of the most common compounds of titanium is titanium dioxide, $TiO_2$, a white solid whose brightness and opacity gives it a wide range of commercial applications. Titanium white pigment was developed in the 1910s and rapidly replaced the previously popular lead white as a much safer alternative. As well as being used by artists, this pigment is found in toothpaste, paper, ink and even in food, and approximately 4.6 million tonnes (5 million tons) of titanium dioxide are used globally as a pigment each year.

Titanium combines strength and lightness, and is highly valued for the manufacture of bicycle frames, such as this De Rosa bicycle, presented at the Dubai Motor Show in 2017.

# Vanadium

DISCOVERY DATE: 1801     DISCOVERED BY: Andrés Manuel del Río

Vanadium takes its name from the Norse goddess Freyja, who has a number of alternative names including Vanadis, the root of this metal's name. The reason behind this choice was that vanadium forms a number of chemical compounds that are highly beautiful, just like the goddess. However, the original discoverer of this bright white, ductile metal had another name in mind for it altogether. Andrés Manuel del Rio, a Spanish scientist working in Mexico, first christened the element panchromium when he discovered it in 1801, because its salts had a wide range of colours. He later changed its name to erythronium, from a Greek root meaning 'red', on realizing that all of these salts turned red when heated or treated with acid. However, when a Swedish chemist, Nils Gabriel Sefström, discovered the same element independently in 1830, he called it vanadium, and the name stuck, despite del Río's protests.

Vanadinite, a mineral with distinctive red crystals, is one of the main ores used in the industrial production of vanadium.

| 23 | |
|---|---|
| **V** | |
| Vanadium | |
| 50.942 | |

| | |
|---|---|
| Atomic number: | 23 |
| Group: | Group 5 |
| Period: | Period 4 |
| Block: | d-block |
| Atomic mass: | 50.942 u |

| | |
|---|---|
| Melting point: | 1910 °C, 3470 °F |
| Boiling point: | 3407 °C, 6165 °F |
| Density: | 6.11 g/cm³ (near room temperature) |
| Appearance: | Blue-grey metal |

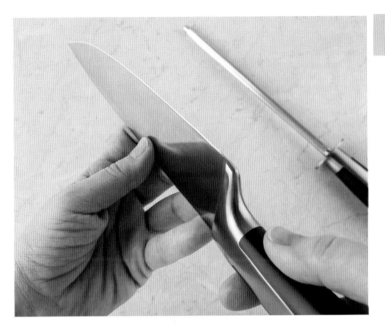

**High-carbon molybdenum-vanadium steel is used for many purposes, including the manufacture of professional chefs' knives.**

From the closing years of the nineteenth century to the present day, vanadium's primary application has been as an additive to steel, where it significantly increases the alloy's strength. For this reason, it is particularly useful in gears, crankshafts, bicycle frames, tools and surgical instruments – all contexts in which corrosion or wear could have disastrous results. Approximately 85 per cent of all vanadium produced globally each year goes into steel production, with 10 per cent being used in titanium alloys and the remaining 5 per cent for other purposes.

# Chromium

DISCOVERY DATE:1797    DISCOVERED BY: Nicolas Louis Vauquelin

Chromium, the first element in Group 6 of the periodic table, is a silver-grey metal that is highly valued for its hardness and its resistance to corrosion. Its name comes from the Greek word *chroma*, meaning 'colour', because many of its compounds have strong colours. The red colour of rubies and the green of emeralds are due to trace amounts of chromium present in these jewels.

Chromium's colourful properties have made it a valuable pigment, due to its brightness and the fact that it does not fade when exposed to daylight. Chrome yellow was chosen as the colour of US school buses and the postboxes of many European countries for this reason – although today this paint is being largely replaced by organic alternatives for environmental and safety-related reasons.

Pure chromium metal is lustrous and corrosion resistant. It is the third-hardest of all the elements, after carbon and boron.

| 24 | |
|---|---|
| **Cr** | |
| Chromium | |
| 51.996 | |

| | |
|---|---|
| Atomic number: | 24 |
| Group: | Group 6 |
| Period: | Period 4 |
| Block: | d-block |
| Atomic mass: | 51.996 u |
| Melting point: | 1907 °C, 3465 °F |
| Boiling point: | 2671 °C, 4840 °F |
| Density: | 7.15 g/cm³ (near room temperature) |
| Appearance: | Hard silvery metal |

Chromium is a vital ingredient in the production of stainless steel, the most common form of which combines iron with 18 per cent chromium and 10 percent nickel (hence the '18/10' mark often found on objects such as stainless steel knives and forks). It is also applied to steel through electroplating as an effective method of preventing rust. This chrome plating is what gives classic cars and motorbikes their shiny glamour.

Human beings require trace amounts of chromium for its role in the action of insulin, and it is also present in RNA. The total amount in a typical human body ranges from 1 to 12 milligrams (0.00003527396–0.000423oz).

This Ford Thunderbird has plenty of chrome detailing to catch the eye of passers-by when it glints in the light.

# Manganese

DISCOVERY DATE: 1774    DISCOVERED BY: Johan Gottlieb Jahn

Manganese is the fifth most abundant metal in the Earth's crust. (This metal, with the chemical symbol Mn, should not be confused with element 12, magnesium, whose chemical symbol is Mg.) The main ore from which manganese is produced, pyrolusite, is primarily composed of manganese dioxide, with the chemical formula $MnO_2$. Today, the synthetically produced version of this compound is used in the production of alkaline and zinc-carbon batteries, but in its natural form it has been known by humankind for millennia. In ancient times it was valued as a blackish-brown pigment, and when the famous cave paintings in Lascaux, France were chemically analyzed, they were found to contain manganese-based pigments.

   In its pure form, manganese is used primarily in the production of steel, with 85 to 90 per cent of the world's manganese production serving this purpose. Standard steel

Manganese is a hard, brittle metal that is silvery-grey in its pure state and tarnishes when exposed to air.

| 25 | |
|---|---|
| **Mn** | |
| Manganese | |
| 54.938 | |

| | |
|---|---|
| Atomic number: | 25 |
| Group: | Group 7 |
| Period: | Period 4 |
| Block: | d-block |
| Atomic mass: | 54.938 u |

| | |
|---|---|
| Melting point: | 1246 °C, 2275 °F |
| Boiling point: | 2061 °C, 3742 °F |
| Density: | 7.21 g/cm³ (near room temperature) |
| Appearance: | Brittle silvery metal |

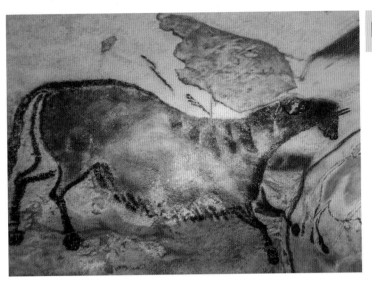

The cave paintings at Lascaux, created 17,000 years ago, contain brown–black manganese pigments.

contains around 1 per cent manganese, where it increases the alloy's strength, while also improving workability and resistance to wear. Steel containing 11 to 15 per cent manganese is extremely strong, and is used in products such as cement mixers, railway crossings, safes and bulletproof cabinets.

Approximately 80 per cent of the world's manganese resources are found in South Africa, with other important deposits in Australia, Brazil, China, Gabon, India and Ukraine. There are also an estimated 500 billion tonnes (551 billion tons) of manganese nodules on the world's ocean floors, but no successful method of harvesting this resource has been found.

All living organisms require manganese, and the human body contains on average around 12 milligrams of the element. It is needed for the functioning of various enzymes and is involved in glucose metabolism. The main sources of manganese in people's diets are cereals and nuts, and other good sources are beetroot, sunflower seeds, olives, avocadoes and tea.

# Iron

DISCOVERY DATE: By at least 3500 BC

Iron is the most abundant element on Earth, and the central core of our planet is believed to be a 2500-km (1553-mile) wide ball of solid iron, which is itself surrounded by vast quantities of molten iron.

The magnetic field generated by all this iron protects us from damaging cosmic radiation, and enables us to navigate the globe by using compasses to show the location of the magnetic North Pole. Many other animals also use the Earth's magnetic currents as a kind of internal 'satnav' to help them in their annual migrations.

The earliest known iron items were discovered in Egypt, dating from around 3500 BCE, and iron was first smelted from ore by the Hittites approximately 2000 years later. During the Industrial Revolution, iron production was expanded rapidly thanks to the invention of coke-fired blast furnaces, which could output huge amounts of

**Haematite, a common compound of iron oxide ($Fe_2O_3$), is an important ore used in the production of iron.**

| 26 | |
|---|---|
| **Fe** | |
| Iron | |
| 55.845 | |

| | |
|---|---|
| **Atomic number:** | 26 |
| **Group:** | Group 8 |
| **Period:** | Period 4 |
| **Block:** | d-block |
| **Atomic mass:** | 55.845 u |

| | |
|---|---|
| **Melting point:** | 1538 °C, 2800 °F |
| **Boiling point:** | 2861 °C, 5182 °F |
| **Density:** | 7.87 g/cm³ (near room temperature) |
| **Appearance:** | Grey metal that rusts to orange when exposed to air |

cast iron. Iron has been crucial in the economic development of humankind for millennia, playing a vital role in the manufacture of weapons, vehicles, tools, bridges and buildings.

One key weakness of iron as a structural material is its tendency to oxidize, or rust, when exposed to oxygen in the air. Strategies to prevent this include galvanizing it with zinc or coating it with tin. Although rarely seen in everyday life, iron that has not been exposed to oxygen has a shiny, silvery surface.

As well as providing strong structures that are vital to human society, we also rely on iron as one of the true building blocks of life. In humans, iron deficiency causes us to produce fewer red blood cells, leading to anaemia, tiredness and breathlessness.

These ornately decorated gates Old Royal Naval College in Greenwich, London, are made from wrought iron.

# Cobalt

DISCOVERY DATE: 1730     DISCOVERED BY: Georg Brandt

Cobalt is a hard, silvery metal and the first element in Group 9 of the periodic table. Its name comes from the German word *Kobold*, meaning 'goblin'. This is because miners had already given the ore that contains cobalt the name 'goblin ore', due to the fact that when smelted, it gave off toxic fumes that contained arsenic – making it a mineral to be avoided when mining for ores containing silver or gold.

Cobalt has been prized since ancient times as a pigment, most famously as the deep, vibrant colour called 'cobalt blue', known to chemists as cobalt(II) aluminate, $CoAl_2O_4$. As well as being used in paintings, glassware and ceramics for aesthetic reasons, it served an additional function in glass bottles, protecting their contents from being damaged by light rays. Other colourful compounds of this element still in use today include cobalt green, cobalt violet, cobalt yellow and cerulean blue (made especially

Cobalt is a hard, shiny metal that forms a protective oxidised coating when exposed to air.

| | |
|---|---|
| **27** | |
| **Co** | |
| Cobalt | |
| 58.933 | |

| | |
|---|---|
| Atomic number: | 27 |
| Group: | Group 9 |
| Period: | Period 5 |
| Block: | d-block |
| Atomic mass: | 58.933 u |

| | |
|---|---|
| Melting point: | 1495 °C, 2723 °F |
| Boiling point: | 2927 °C, 5301 °F |
| Density: | 8.90 g/cm³ (near room temperature) |
| Appearance: | Hard blue-grey metal |

famous by Meryl Streep's speech in the film *The Devil Wears Prada*). Other commercial applications of cobalt are in the production of lithium-ion batteries, high-strength metal alloys, and in electroplating.

All animals require cobalt because it is part of cobalamin, or vitamin B12. This vitamin supports the functioning of the nervous system, the production of red blood cells and the metabolism of food. It is also involved in the synthesis of DNA. The best sources of vitamin B12 for humans are animal-derived foods such as meat, chicken, fish, eggs and dairy products, and fortified grain-based foods.

Cobalt has been used since antiquity to provide a vivid blue colour to ceramics, jewellery, paint and glassware such as these wine glasses.

# Nickel

DISCOVERY DATE: 1751     DISCOVERED BY: Axel Fredrik Cronstedt

Nickel is a hard, silvery metal with a slight gold tinge to it. Its name comes from an ore called *kupfernickel* in German, meaning 'devil's copper' or 'goblin copper' because it looked like a potentially copper-bearing mineral, but did not yield any copper. Its origin story is therefore similar to that of its neighbour in the table, cobalt, with a scientist – in this case, Swedish chemist Axel Fredrik Cronstedt – isolating a new element from a maligned mineral.

Nickel has been known since ancient times, with the earliest recorded use dating to 3500 bce. Its main applications today are in metal alloys, with about 68 per cent of global production being used in stainless steel. Other key uses are in electroplating, rechargeable batteries and coinage. The American 5-cent coin is called a nickel, although it's in fact made from an alloy of 25 per cent nickel and 75 per cent copper.

Nickel, a shiny silvery-white metal with a yellow-gold tinge, has been used for thousands of years and appears today in countless products including batteries, guitar strings, steel and other alloys.

| 28 | |
|---|---|
| **Ni** | |
| Nickel | |
| 58.693 | |

| Atomic number: | 28 |
|---|---|
| Group: | Group 10 |
| Period: | Period 4 |
| Block: | d-block |
| Atomic mass: | 58.693 u |

| Melting point: | 1455 °C, 2651 °F |
|---|---|
| Boiling point: | 2730 °C, 4946 °F |
| Density: | 8.908 g/cm³ (near room temperature) |
| Appearance: | Silvery metal with a gold tinge |

**Ni**

Nickel deposits were discovered by French colonists in New Caledonia in 1864, and the archipelago contains 25 per cent of the world's nickel deposits.

Nickel is essential for many forms of life, including plants, fungi, bacteria and archaea (single-celled organisms), but it has not been officially confirmed as a vital element for humans, and there is no recommended daily amount from the US Institute of Medicine or the UK's National Health Service. It is estimated that the average person consumes 70-100 micrograms (2.46–3.52oz), and that we absorb only 10 per cent of this amount. Foods rich in nickel are legumes, chocolate, oats, soy beans, seeds and nuts, although exact levels of nickel in any food will vary depending on the soil it was grown in.

# Copper

DISCOVERY DATE: Prehistoric

Copper is a malleable and ductile metal with an orange-pink colour that changes to a pale green when it oxidizes on prolonged exposure to the air (as is often seen on church spires and weathervanes). It has been known for millennia, with evidence of human use dating back more than 10,000 years. Its name comes from the fact that the principal source of copper in the Roman era was Cyprus: the Latin *aes Cyprium* became *cuprum*, eventually giving us the English word 'copper'. Today, the largest deposits of copper ore are found in Canada, Chile, Peru, the USA, Zaire and Zambia.

Copper is an excellent conductor of electricity – silver is the only element with a higher conductivity – and so it is widely employed in the manufacture of electrical equipment and wiring. It is also used in metal alloys, and in fact, the first ever alloy made by humans was a mixture of copper and tin. The resulting metal was strong,

Copper can be found in a wide variety of ores, which are processed to produce pure copper for industrial applications.

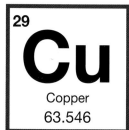

**29**

# Cu

Copper

63.546

| | |
|---|---|
| Atomic number: | 29 |
| Group: | Group 11 |
| Period: | Period 4 |
| Block: | d-block |
| Atomic mass: | 63.546 u |

| | |
|---|---|
| Melting point: | 1084.62 °C, 1984.32 °F |
| Boiling point: | 2562 °C, 4643 °F |
| Density: | 8.96 g/cm³ (near room temperature) |
| Appearance: | Red-orange metal |

hard and easy to shape, and was perfect for making tools and weapons. It was so important that it gave its name to an entire era: the Bronze Age. Later, the Romans alloyed copper with zinc to create brass, another metal still much in use today.

One rare characteristic of copper is that it can be recycled without any loss of quality. Amazingly, it is estimated that 80 per cent of the copper that has ever been mined is still in use.

Copper is essential for all species of life for its role in certain enzymes, and the average human body contains about 70 milligrams (0.00246oz). In many invertebrates, it fulfils the same role in human blood as iron, transporting oxygen around the body. This means that, instead of having red blood like us, animals such as octopuses, butterflies, snails and oysters have pale blue blood.

A domestic cooking pan made with copper, it is famed for its ability to conduct heat and electricity. However, copper is a reactive metal and it's best to used lined pans to avoid them leaching copper into food during the cooking process, making them unsafe to use.

# Zinc

DISCOVERY DATE: Antiquity    DISCOVERED BY: Isolated in its pure form by Andreas Marggraf in 1746

Zinc is a silvery metal with a bluish tinge that tarnishes when exposed to air. Its name comes from the German word *Zinke*, meaning 'prong'; this is because it takes the form of sharp pointed crystals after smelting.

In the human body, zinc is vital for the function of more than 300 enzymes and 1000 transcription factors. For this reason, its roles are too numerous to list here, but it is involved in functions such as the immune system, metabolism, healing of wounds and the senses of taste and smell. It plays a key role in the liver for anyone who enjoys a glass of wine, beer or spirits, as part of the alcohol dehydrogenase enzyme; this enzyme helps undo the toxic effects of some of our favourite tipples. Good sources of dietary zinc include red meat, chicken, fortified cereals, nuts, peas and seeds. An estimated two billion people in the world suffer from zinc deficiency, which is

Sphalerite, or zinc oxide, is the most important ore of zinc, accounting for some 95 per cent of all zinc production.

| | |
|---|---|
| **30** | |
| **Zn** | |
| Zinc | |
| 65.38 | |

| | |
|---|---|
| **Atomic number:** | 30 |
| **Group:** | Group 12 |
| **Period:** | Period 4 |
| **Block:** | d-block |
| **Atomic mass:** | 65.38 u |

| | |
|---|---|
| **Melting point:** | 419.53 °C, 787.15 °F |
| **Boiling point:** | 907 °C, 1665 °F |
| **Density:** | 7.14 g/cm³ (near room temperature) |
| **Appearance:** | Silver-white metal |

associated with increased susceptibility to infection, diarrhoea, malnutrition, liver and kidney problems and diabetes.

Zinc has been used since ancient times, with finds of Judean brass (an alloy of copper with 23 per cent zinc) dating as far back as the fourteenth century BCE. The main applications of zinc today are as a coating for galvanized iron, in batteries and in metal alloys. Compounds containing zinc are used in a wide range of applications, including deodorants, dietary supplements, anti-dandruff shampoos, luminescent paints, make-up and sunblock.

Oysters are a good source of dietary zinc, along with meat, poultry, other seafoods, legumes and whole grains.

# Gallium

DISCOVERY DATE:1875     DISCOVERED BY: Paul-Émile Lecoq de Boisbaudran

When compiling his periodic table of the elements, Dmitri Mendeleev predicted the existence of gallium, even though it had not been discovered yet. He called it 'eka-aluminium', foreseeing that it was likely to have similar characteristics to the metal above it in Group 13, aluminium.

Gallium is a soft silvery metal that becomes silvery-white in its liquid form. It was discovered by French chemist Paul-Émile Lecoq de Boisbaudran in 1875, and he based its name on the Latin word for France, Gallia. Some of his contemporaries wondered whether he was making a pun on the Latin for 'rooster', *gallus*, because this would be *le coq* in French, like his own surname, but he denied this. (To date, no scientist has ever named an element after themselves – but there is a first time for everything.)

Pure gallium is a soft silver metal that melts at temperatures just above standard room temperature.

**31**

# Ga

Gallium
69.723

| Atomic number: | 31 |
|---|---|
| Group: | Group 13 |
| Period: | Period 4 |
| Block: | p-block |
| Atomic mass: | 69.723 u |

| Melting point: | 29.7646 °C, 85.5763 °F |
|---|---|
| Boiling point: | 2403 °C, 4357 °F |
| Density: | 5.91 g/cm³ (near room temperature) |
| Appearance: | Silvery-blue metal |

Gallium has no known biological role, but clinical trials have suggested that gallium nitrates could be effective as a treatment for some forms of cancer, and they are also used in nuclear medicine imaging, known as a gallium scan.

In industry, gallium has been designated as a technology-critical element, with 98 per cent of its use going into semiconductors. It is needed in a huge range of items we rely on today, such as solar panels, blue and green LEDs (light-emitting diodes), integrated circuits and mobile phones.

**If you hold a piece of solid gallium in your hand for a few minutes, it will become liquid, because its melting point of 29.76 °C is lower than human body temperature.**

# Germanium

DISCOVERY DATE:1886     DISCOVERED BY: Clemens Winkler

Like gallium, the previous element in the periodic table, the existence of germanium was predicted by Dmitri Mendeleev long before it was discovered; he gave it the placeholder name of 'eka-silicon', the element directly above it in Group 14. Mendeleev had predicted that the new element would have an atomic weight of approximately 72.64, and when Clemens Winkler discovered it in 1886, he found its weight to be 72.59 – which was pretty impressive. Its name comes from Winkler's home country of Germany.

Germanium shares many characteristics with its 'upstairs' and 'downstairs' group neighbours, silicon and tin, and visually it resembles silicon, being shiny, brittle and silvery-white in colour. It is an electrical semiconductor, and was the only element used for this purpose in the first decade of the development of semiconductor

Germanium is a shiny, brittle metal that was first discovered in 1886.

| 32 | |
|---|---|
| **Ge** | |
| Germanium | |
| 72.630 | |

| Atomic number: | 32 |
|---|---|
| Group: | Group 14 |
| Period: | Period 4 |
| Block: | p-block |
| Atomic mass: | 72.630 u |

| Melting point: | 938.25 °C, 1720.85 °F |
|---|---|
| Boiling point: | 2833 °C, 5131 °F |
| Density: | 5.323 g/cm³ (near room temperature) |
| Appearance: | Silvery-white semi-metal |

technology, although today other elements are preferred in this role. Because it has a high refractive index, germanium has a lot of applications related to light: it is used in lenses for microscopes and wide-angle cameras, in fibreoptics, LEDs (light-emitting diodes) and in solar cells. The industrial production of germanium is based upon its extraction from zinc, silver, lead and copper ores, and from the ash produced by coal-fuelled power plants.

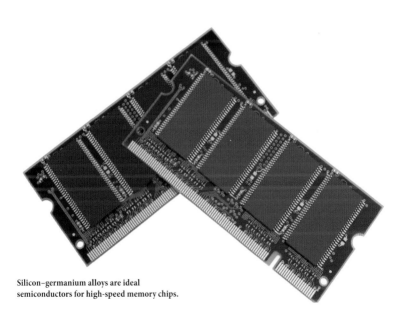

**Silicon–germanium alloys are ideal
semiconductors for high-speed memory chips.**

# Arsenic

DISCOVERY DATE: Approximately 1250 CE          DISCOVERED BY: Albertus Magnus

If there were a popularity league table for the elements, it's likely that gold and silver would take the top spots, and a selection of the more famously dangerous elements (plutonium, uranium) would be bumping along the bottom. However, the one that would almost certainly get the biggest skull-and-crossbones warning signs in a public survey is element 33: arsenic.

The toxic properties of this element, which can be found in grey, yellow and black allotropes, have been known since ancient times. In 370 BCE, the Greek physician Hippocrates recognized it as a poison but also recorded that it could be used as a treatment for ulcers. Its popularity as a medicine continued well into the Victorian era, in a popular tonic called Fowler's Solution, and as a beauty treatment to achieve a pale complexion. At the same time, it was ideal as a murder weapon, because it was easy to

Arsenic has a grey metallic appearance and is notoriously poisonous.

| 33 | |
|---|---|
| **As** | |
| Arsenic | |
| 74.922 | |

| | |
|---|---|
| Atomic number: | 33 |
| Group: | Group 15 |
| Period: | Period 4 |
| Block: | p-block |
| Atomic mass: | 74.922 u |

| | |
|---|---|
| Melting/ boiling point: | 616°C, 1141°F (sublimes directly from solid to gas) |
| Density: | 5.727 g/cm³ (near room temperature) |
| Appearance: | Silver-grey semi-metal |

The Chinese brake fern, *Pteris vittata*, is native to temperate regions of Eurasia, Africa and Australia. It hyperaccumulates arsenic into its leaves from soil and could be used as a phytoremediation measure to remove arsenic from contaminated land.

obtain, had no strong flavour and caused a broad range of side effects as it built up in a victim's body, meaning that it might successfully avoid raising suspicion.

When not being consumed (voluntarily or not), arsenic found other roles in Victorian homes, as a preservative in lace, a green pigment, an insecticide and an anti-mould treatment in wallpaper. In the last of these functions, it is even thought to have played a part in the death of Napoleon I in 1821 during his island exile on St Helena.

Despite its poor reputation, arsenic still has a number of industrial applications, the main one of which is as an alloying agent with lead, where a very small amount of arsenic adds strength to the compound. It is also used in bronzing and pyrotechnics, and in the production of semiconductors.

# Selenium

DISCOVERY DATE: 1817     DISCOVERED BY: Jöns Jacob Berzelius

The story behind the naming of selenium will appeal to lovers of the classics and astronomy. It was named after Selene, the Greek goddess of the Moon, because it is chemically similar to the previously discovered element tellurium, whose name comes from the Latin *tellus*, meaning 'earth'.

Selenium forms several different allotropes, including a red amorphous powder, black vitreous beads and a grey crystalline metallic lattice. About half of the global production of selenium is used in the manufacture of glass; it is particularly valued because it adds a red colour that cancels out any yellow or green tints, which are common in glass due to iron impurities. Its electrical resistance varies according to the amount of light shining on it, making it useful in photocopiers, camera light meters, solar cells and electric eyes.

Selenium has several allotropes – different physical forms of the same element. This grey form of selenium is a semiconductor and is resistant to oxidation by air.

| 34 | | |
|---|---|---|
| **Se** | | |
| Selenium | | |
| 78.971 | | |

| Atomic number: | 34 |
|---|---|
| Group: | Group 16 |
| Period: | Period 4 |
| Block: | p-block |
| Atomic mass: | 78.971 u |

| Melting point: | 221 °C, 430 °F |
|---|---|
| Boiling point: | 685 °C, 1265 °F |
| Density: | 4.81 g/cm³ |
| | (grey form, near room temperature) |
| Appearance: | Silvery-grey metal or red powder |

Selenium is a vital trace element that can be found in meat, nuts, mushrooms and cereals, or taken as a dietary supplement. Brazil nuts contain a large amount of selenium.

For animals, selenium is an essential trace element, and in humans it supports the immune system and helps to prevent damage to cells and tissues. However, taking in too much selenium can cause harmful side effects, such as the loss of hair and nails, bad breath, tiredness and irritability. This toxicity gives selenium a commercial application as the active ingredient in many anti-dandruff shampoos, because it is toxic to the scalp fungus that causes dandruff.

# Bromine

DISCOVERY DATE: 1826 DISCOVERED BY: Antoine-Jérôme Balard (France) and Carl Jacob Löwig (Germany, 1825)

Bromine's name does not beat about the bush when it comes to describing this element's key characteristic, coming from the Greek word *bromos*, meaning 'stench'. It was discovered independently by Carl Jacob Löwig in Germany and Antoine-Jérôme Balard in France only a year apart; although Löwig isolated the element in 1825, a year before Balard, Balard published his findings first, and was thus named officially as its discoverer.

At room temperature, bromine is a red-brown liquid, one of only two elements to be liquid at room temperature, the other being mercury. It has a sharp, pungent smell, as heralded by its name, and it evaporates easily into a similarly coloured gas. Because it is highly reactive, it is not found in its pure form in nature; instead, it is evaporated from bromine-rich brine ponds in China, Israel and the USA.

At room temperature, bromine is a red-brown liquid that readily evaporates into a similarly coloured vapour.

| 35 | |
|---|---|
| **Br** | |
| Bromine | |
| 79.904 | |

| | |
|---|---|
| Atomic number: | 35 |
| Group: | Group 17 |
| Period: | Period 4 |
| Block: | p-block |
| Atomic mass: | 79.904 u |

| | |
|---|---|
| Melting point: | −7.2 °C, 19 °F |
| Boiling point: | 58.8 °C, 137.8 °F |
| Density: | 3.1028 g/cm³ (near room temperature) |
| Appearance: | Red-brown liquid |

When heated, organobromine compounds produce free bromine atoms, in a process that prevents free radical chemical chain reactions. As a result of this property, these compounds are effective fire retardants, and they are added to foam fillings for furniture and plastic casings for electronics as a safety measure. Bromine is also used to sanitize water in pools and spas, where it can be preferred to chlorine because of its milder smell.

Bromine in the form of bromide is present in trace amounts in the human body but has no known biological function. Elemental bromine is toxic to humans, irritating the eyes, nose and throat and causing burns to the skin.

The water of the Dead Sea contains a concentration of 4500–5000 ppm of bromine, and Israel and Jordan are the leading production sites for this element.

# Krypton

DISCOVERY DATE: 1898     DISCOVERED BY: William Ramsay and Morris Travers

Krypton is a chemically inert, colourless, odourless and tasteless gas, and therefore a classic member of Group 18, the noble gases. Its name comes from the Greek word kryptos, meaning 'the hidden one'. It was discovered in 1898 by the British chemists Sir William Ramsay and Morris Travers, who were evaporating the components of liquid air in the hope of finding new elements. Their discovery of krypton was followed a few weeks later by the isolation of neon using the same process, and William Ramsay was awarded the 1904 Nobel Prize in Chemistry for his achievements in the discovery of a number of noble gases.

Krypton is used in photographic flashbulbs, where it gives a bright white light, and it is also used in combination with mercury to create luminous displays with a bright blue-green light. Because its light is so bright, krypton is used in the high-powered

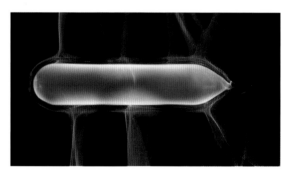

Ionised krypton gas gives off a characteristic blue-white glow.

| 36 | |
|---|---|
| **Kr** | |
| Krypton | |
| 83.798 | |

| | |
|---|---|
| **Atomic number:** | 36 |
| **Group:** | Group 18 |
| **Period:** | Period 4 |
| **Block:** | p-block |
| **Atomic mass:** | 83.798 u |
| **Melting point:** | −157.37 °C, −251.27 °F |
| **Boiling point:** | −153.415 °C, −244.147 °F |
| **Density:** | 0.00375 g/cm³ (near room temperature) |
| **Appearance:** | Colourless gas |

**Krypton lamps provide a strong bright light to guide aeroplanes safely down onto airport runways.**

lamps on airport runways. Krypton has no known biological role and is non-toxic. However, when breathed in, it has an anaesthetic effect, which could be dangerous in the event of a leak.

Between 1960 and 1983, krypton was used to define the internationally agreed length of a metre, which was deemed to be identical to the wavelength of light emitted from the krypton isotope Kr-86. (In 1983, this definition was replaced by the distance that light travels in vacuum during 1/299,792,458 seconds.)

# Rubidium

DISCOVERY DATE: 1861    DISCOVERED BY: Gustav Kirchhoff and Robert Bunsen

Rubidium, a soft, silvery-white metal, was discovered by the chemists Robert Bunsen and Gustav Kirchhoff in Germany in 1861, using a new technique called flame spectroscopy. They dissolved the mineral lepidolite, which was known to contain lithium and potassium, and then examined the atomic spectrum of what remained after the other elements had been removed. Seeing two intense red lines in the emission spectrum of the newly isolated element, they named it after the Latin word *rubidus*, meaning 'deep red'.

Like other members of Group 1, rubidium is highly reactive; it reacts violently when added to water, and can even ignite spontaneously on exposure to the air. To prevent these outcomes, rubidium is stored under a protective layer of oil or grease.

This glass ampoule contains a sample of rubidium, a metal so reactive that it must be stored away from air and water.

| 37 | Rb | Rubidium | 85.468 |
|---|---|---|---|

| | |
|---|---|
| **Atomic number:** | 37 |
| **Group:** | Group 1 |
| **Period:** | Period 5 |
| **Block:** | s-block |
| **Atomic mass:** | 85.468 u |

| | |
|---|---|
| **Melting point:** | 39.30 °C, 102.74 °F |
| **Boiling point:** | 688 °C, 1270 °F |
| **Density:** | 1.532 g/cm³ (near room temperature) |
| **Appearance:** | Silvery-white metal |

**Rb**

Rubidium gives a striking purple colour to these exploding fireworks.

Rubidium is not as widely produced or used as many of the other elements we have encountered so far in this book, and one of its primary applications is in research. It is used in photocells, to make special types of glass, and as a 'getter' to remove trace gases from vacuum tubes. Like caesium, the element immediately below it, rubidium is used in atomic clocks. Outside of the laboratory, it also plays a role in fireworks, where it gives a purple flame to the explosions.

Rubidium has no known biological role, but it can become concentrated in human cells due to its similarity to potassium. This characteristic makes it useful in medical imaging, especially in scans for brain tumours.

# Strontium

DISCOVERY DATE: 1790     DISCOVERED BY: Adair Crawford

Strontium is a soft silver-white metal that reacts vigorously with water, and that will spontaneously burst into flame at room temperature when finely powdered. It is the fifteenth most abundant element on Earth, but it is not found in a pure form due to its propensity to form compounds with other elements (which is probably lucky for anyone who might stumble upon it). It usually occurs in the minerals strontianite and celestite, and its name comes from Strontian, the Scottish village three miles south of the lead mines where it was first identified.

Strontium is used in the fireworks industry for its brilliant red light, and the compound strontium aluminate is the active ingredient in plastics and paints that glow in the dark, absorbing light during the day and emitting it for several hours afterwards. It was once much in demand for the glass of cathode-ray tube (CRT)

Celestite, named for its clear pale blue colour, is a mineral form of strontium sulfate, and a key source of elemental strontium.

| 38 | |
|---|---|
| **Sr** | |
| Strontium | |
| 87.62 | |

| | |
|---|---|
| **Atomic number:** | 38 |
| **Group:** | Group 2 |
| **Period:** | Period 5 |
| **Block:** | s-block |
| **Atomic mass:** | 87.62 u |

| | |
|---|---|
| **Melting point:** | 777 °C, 1431 °F |
| **Boiling point:** | 1377 °C, 2511 °F |
| **Density:** | 2.64 g/cm³ (near room temperature) |
| **Appearance:** | Soft silvery metal |

televisions, because it prevents X-ray emissions while allowing light through without any discolouration. This use once accounted for three-quarters of the total output of strontium, but since the widespread replacement of CRT televisions by flatscreen technology, this demand has declined significantly.

One of strontium's isotopes, strontium-90, is infamous as a radiation hazard present in the fallout from nuclear tests and accidents. It has a half-life of 29 years and thus takes centuries to fall to negligible levels, and its tendency to accumulate in bones and bone marrow means that it can cause bone cancer and leukaemia. However, in controlled amounts, this very same isotope can be effective in radiotherapy and as a radioactive tracer.

Radioactive strontium-90 was present in the fallout from the disaster at the nuclear power plant at Chernobyl, Ukraine. This photograph was taken a few weeks after the disaster in May 1986.

# Yttrium

DISCOVERY DATE: 1794    DISCOVERED BY: Johan Gadolin

Yttrium was named after Ytterby, a village in Sweden, close to where it was first found. The same village also gave its name to three of the lanthanide group elements that we'll encounter later in this book: terbium (65), erbium (68) and ytterbium (70). The Finnish chemist Johan Gadolin discovered yttrium oxide, $Y_2O_3$, in a mineral called ytterbite, which was later renamed gadolinite in his honour. Pure yttrium, a soft, silver-grey metal with a crystalline form, was finally isolated in 1828 by the German chemist Friedrich Wöhler.

Yttrium has no biological role, but trace amounts of it are found in all organisms, with the average human body containing about 0.5 milligrams (0.0000176oz). If inhaled, yttrium causes permanent damage to the lungs and liver. However, the radioactive isotope yttrium-90 is valued as a treatment for certain cancers,

Yttrium is a soft silvery-white metal with a highly crystalline structure.

**39**

**Y**

Yttrium

88.906

| | |
|---|---|
| Atomic number: | 39 |
| Group: | Group 3 |
| Period: | Period 5 |
| Block: | d-block |
| Atomic mass: | 88.906 u |

| | |
|---|---|
| Melting point: | 1526 °C, 2779 °F |
| Boiling point: | 2930 °C, 5306 °F |
| Density: | 4.472 g/cm³ (near room temperature) |
| Appearance: | Soft silvery metal |

including leukaemia, lymphoma, bone, ovarian, colorectal and pancreatic. There are a wide range of industrial applications of yttrium. It is used in alloys to strengthen magnesium and aluminium, and in the production of synthetic garnets, which are not only used for jewellery, but also in lasers for medical procedures such as skin resurfacing, dentistry and ophthalmology. Yttirum is also found in other contexts, such as in the electrodes of high-performance spark plugs, in cathode-ray tube (CRT) screens, where it combines with europium to display the colour red, and as an electrical superconductor.

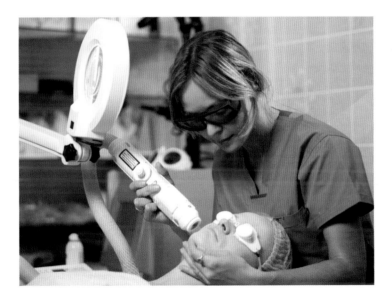

Neodymium-doped yttrium aluminium garnet (Nd:YAG) lasers are used to treat a variety of skin problems such as thread veins and vascular birthmarks.

# Zirconium

DISCOVERY DATE: 1789    DISCOVERED BY: Martin Heinrich Klaproth

Zirconium is a hard, silvery-grey metal in Group 4 of the periodic table. Its name comes from the Persian word *zargun*, meaning 'gold-coloured', which may seem odd for a silver-coloured metal, but this name was originally attached to its parent ore, zircon, which is found in a range of warm golden colours, from yellow through to orange and red.

The main application of zirconium today is in nuclear reactors, where the fact that it does not absorb neutrons makes it perfect for use in tubing. As a single nuclear reactor could have more than 100,000m (328,000ft) of zirconium alloy tubing, it's no surprise that this application accounts for 90 per cent of the use of this element. Zirconium metal is also used in jet engines, gas turbines and spacecraft, where its high resistance to heat is especially needed. Like titanium, zirconium forms a thin layer of oxidation

Zirconium is a shiny, strong transition metal that is ductile, malleable and solid at room temperature.

| 40 | |
|---|---|
| **Zr** | |
| Zirconium | |
| 91.224 | |

| Atomic number: | 40 |
|---|---|
| Group: | Group 4 |
| Period: | Period 5 |
| Block: | d-block |
| Atomic mass: | 91.224 u |

| Melting point: | 1852 °C, 3365 °F |
|---|---|
| Boiling point: | 4377 °C, 7911 °F |
| Density: | 6.52 g/cm³ (near room temperature) |
| Appearance: | Silvery-white metal |

on its outer surface, making it resistant to corrosion from acids, alkalis and seawater, and therefore it is used for pipes, fittings and heat exchangers where this kind of corrosion is likely.

A more glamorous application of zirconium is in the compound $ZrO_2$, or cubic zirconia, which looks practically identical to diamond. Being lab-produced rather than mined, it is much cheaper than diamond, and only an expert eye can spot the difference.

Cubic zirconia, a crystalline form of zirconium dioxide, $ZrO_2$, is visually almost indistinguishable from diamond and is a much more affordable was to add sparkle to your costume.

# Niobium

DISCOVERY DATE: 1801    DISCOVERED BY: Charles Hatchett

Niobium is a soft, ductile grey metal with a relatively low density. It was discovered in 1801 in a mineral sample from Connecticut held at the British Museum by an English chemist, Charles Hatchett, who gave it the name 'columbium' in honour of America. In 1844, he German chemist Heinrich Rose renamed the element niobium after the Greek goddess Niobe. His logic was that niobium is hard to distinguish from tantalum, which lies immediately below it in Group 5 of the periodic table, and that Niobe was the daughter of Tantalus. Both names remained in use until the International Union of Pure and Applied Chemistry decided in 1949 that niobium would be the official name. However, despite the formal name change, some have refused to let go of tradition, and the United States Geological Survey is one organization that still uses 'columbium' to refer to this element.

Niobium is not the most well-known element, but it plays a vital role in electronic devices, special steel alloys and the superconducting magnets in MRI scanners.

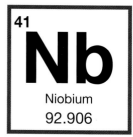

**41**

**Nb**

Niobium

92.906

| | |
|---|---|
| Atomic number: | 41 |
| Group: | Group 5 |
| Period: | Period 5 |
| Block: | d-block |
| Atomic mass: | 92.906 u |

| | |
|---|---|
| Melting point: | 2477 °C, 4491 °F |
| Boiling point: | 4744 °C, 8571 °F |
| Density: | 8.57 g/cm³ (near room temperature) |
| Appearance: | Silvery metal |

The largest application of niobium is in steel alloys used in cars, aircraft and pipelines. It is also used in superalloys with nickel, cobalt or iron in contexts where its heat resistance is especially valuable, such as in jet engines, gas turbines, turbochargers and combustion equipment.

Like its next-door neighbour in the table, zirconium, niobium is used in jewellery making. Not only is it hypoallergenic, but when heated and anodized, it creates a wide range of eye-catching iridescent colours. This property has also seen niobium used for commemorative coinage in a number of countries. For the past 20 years, the Austrian Mint has issued a special 25-euro coin decorated with a colourful two-tone niobium design featuring a current topic from science and technology. The last four designs highlighted big data, smart mobility, extraterrestrial life and global heating.

Key niobium deposits are found in Australia, Brazil, Canada, Democratic Republic of Congo, Mozambique, Nigeria and Rwanda.

# Molybdenum

DISCOVERY DATE: 1781    DISCOVERED BY: Peter Jacob Hjelm

Molybdenum is the 58th most abundant element in the Earth's crust and the 25th most abundant in our oceans. It is a fairly soft silvery-white metal and is one of the best thermal conductors of all the elements. Its name comes from the Greek word *molybdos*, meaning 'lead', because the mineral that it was discovered in, molybdenite, was originally thought to be a lead ore.

Global reserves of molybdenum are thought to be around 19 million tonnes (20.9 million tons), primarily located in China, the USA and South America, and around 190,000 tonnes (209,000 tons) are produced annually. Although molybdenum is quite brittle in its pure form, it adds strength and heat resistance to metal alloys, and 'moly steel' containing between 0.25 and 8 per cent molybdenum accounts for some three-quarters of the total demand for this element. Because of its high melting point,

Molybdenum occurs naturally on Earth within minerals such as molybdenite ($MoS_2$). In its pure form, molybdenum is a silvery-grey metal.

| 42 | |
|---|---|
| **Mo** | |
| Molybdenum | |
| 95.95 | |

| | |
|---|---|
| Atomic number: | 42 |
| Group: | Group 6 |
| Period: | Period 5 |
| Block: | d-block |
| Atomic mass: | 95.95 u |

| | |
|---|---|
| Melting point: | 2623 °C, 4753 °F |
| Boiling point: | 4639 °C, 8382 °F |
| Density: | 10.28 g/cm³ (near room temperature) |
| Appearance: | Silvery-white metal |

molybdenum is used for the electrodes in glass furnaces and for some electrical filaments.

There are 39 known isotopes of molybdenum. One of these, molybdenum-99, is produced as a parent isotope to technetium-99, which is used in medical imaging in approximately 40 million procedures per year.

In the human body, molybdenum is involved in the production and function of enzymes that repair and build genetic material. Good dietary sources include dairy products, wholegrains, bananas, leafy vegetables, beef, chicken and eggs.

**Molybdenum steel plating was used to provide protection for some British tanks in the First World War.**

# Technetium

DISCOVERY DATE: 1937     DISCOVERED BY: Carlo Perrier and Emilio Segrè

Technetium, element number 43, was the first element to be produced artificially, instead of being isolated from a pre-existing compound. Its name comes from the Greek word *technetos*, meaning 'craft' or 'art', therefore signifying the fact that it was created in a laboratory. The existence of technetium had long been a mystery to researchers, as Dmitri Mendeleev had predicted its characteristics and given it the placeholder name 'eka-manganese', reflecting its position just below this element in the periodic table, but nobody could find it. The reason that technetium did not exist naturally on Earth is that it is radioactive, and its longest-lived isotope has a half-life of only 4.2 million years. While this might seem like a long time when viewed from our human perspective, what it means is that any technetium present when the Earth was formed would have decayed away to nothing long ago.

The first generator of technetium-99m was used to produce a metastable isotope of technetium from a decaying sample of molybdenum-99.

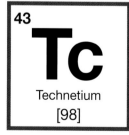

| 43 | |
|---|---|
| **Tc** | |
| Technetium | |
| [98] | |

| | |
|---|---|
| Atomic number: | 43 |
| Group: | Group 7 |
| Period: | Period 5 |
| Block: | d-block |
| Atomic mass: | [98] |
| Melting point: | 2157 °C, 3915 °F |
| Boiling point: | 4265 °C, 7709 °F |
| Density: | 11 g/cm³ (near room temperature) |
| Appearance: | Radioactive silvery metal |

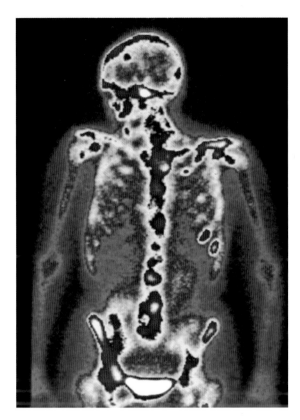

Technetium-99m is used as a radioactive tracer in medical imaging equipment. This nuclear medical bone scan shows multiple focal metastatic lesions on the lungs.

Although three German chemists, Walter Noddack, Otto Berg and Ida Tacke, had claimed the discovery of element 43 in 1925, calling it masurium, their results could not be replicated in other laboratories, and their finding was dismissed. Then in 1937, Carlo Perrier and Emilio Segrè of the University of Palermo in Italy bombarded molybdenum atoms with deuterons that had been accelerated in Ernest Lawrence's cyclotron at the University of California, Berkeley, USA, and succeeded in identifying two isotopes of technetium.

Today, the isotope technetium-99 is used in medical tests, and it can also be used in metal alloys as a corrosion inhibitor. However, because of its high radioactivity, its use for this purpose is restricted to sealed systems, and its range of applications is limited.

# Ruthenium

DISCOVERY DATE: 1844    DISCOVERED BY: Karl Ernst Claus

Ruthenium is a shiny silvery-white metal in the platinum group of the periodic table (made up of ruthenium, rhodium, palladium, osmium, iridium and platinum). It was named after Ruthenia, a region covering today's western Russia, Belarus and Ukraine along with parts of Poland and Slovakia. One of the rarest metals on Earth, ruthenium is the 78th most abundant metal in the Earth's crust, and around 30 tonnes (29.5 tons) are mined each year from an estimated total global reserve of 5000 tonnes (4921 tons).

Ruthenium does not tarnish at room temperature, and it is resistant to damage by air, water and acids, including aqua regia, a mixture of nitric acid and hydrochloric acid that can dissolve gold and silver. It can be alloyed with platinum and palladium to increase the hardness of those metals, and titanium becomes 100 times more resistant to corrosion with the inclusion of only 0.1 per cent ruthenium. Another

Ruthenium is incredibly rare but has a wide range of applications across electronics, electrochemistry, superalloys for jet engines and even in fountain pen nibs.

| 44 | | |
|---|---|---|
| **Ru** | | |
| Ruthenium | | |
| 101.7 | | |

| | |
|---|---|
| **Atomic number:** | 44 |
| **Group:** | Group 8 |
| **Period:** | Period 5 |
| **Block:** | d-block |
| **Atomic mass:** | 101.7 u |

| | |
|---|---|
| **Melting point:** | 2334 °C, 4233 °F |
| **Boiling point:** | 4150 °C, 7502 °F |
| **Density:** | 12.45 g/cm³ (near room temperature) |
| **Appearance:** | Silvery-white metal |

key application of ruthenium is in the electronics industry, where it is used for chip resistors and electrical contacts.

For anyone wishing to add a touch of luxury to their daily writing tasks, a fountain pen with a ruthenium-plated grip or nib would be just the thing. The benefits of such an exclusive pen are largely aesthetic, but the ruthenium would help to protect it from corrosion. Ruthenium is also used as a plating metal for jewellery, where it adds a dark, pewter-like lustre that is hard-wearing and scratch-resistant.

**Ruthenium-based compounds that absorb light can be used in dye-sensitised solar cells, and have the potential to provide low-cost solar power.**

Transition metals

# Rhodium

DISCOVERY DATE: 1804     DISCOVERED BY: William Hyde Wollaston

Rhodium is one of the rarest metals on Earth, and the 79th most abundant element in the Earth's crust. It is shiny, extremely hard and resists corrosion, and its name comes from the Greek word *rhodon*, meaning 'rose', because the crystals of rhodium sodium chloride from which it was first obtained were rose-red in colour.

   The main application of rhodium, accounting for 80 per cent of its annual use, is in the three-way catalytic converters found in all modern cars. These devices reduce pollution by converting carbon monoxide to carbon dioxide and hydrocarbons to carbon dioxide and water through oxidation, and by converting nitrogen oxides to nitrogen and oxygen gases through reduction. Other uses for rhodium are as a catalyst in the chemicals industry, as a coating for optical fibres and mirrors, and as an electrical contact material.

Although rhodium's name comes from the Greek word for 'rose' due to the pink colour of its salts, pure rhodium has a lustrous silver colour.

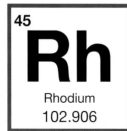

**45**

**Rh**

Rhodium

102.906

| | |
|---|---|
| Atomic number: | 45 |
| Group: | Group 9 |
| Period: | Period 5 |
| Block: | d-block |
| Atomic mass: | 102.906 u |

| | |
|---|---|
| Melting point: | 1964 °C, 3567 °F |
| Boiling point: | 3695 °C, 6683 °F |
| Density: | 12.41 g/cm³ (near room temperature) |
| Appearance: | Silvery-white metal |

Like ruthenium, rhodium is also used in jewellery making, where it gives a bright and highly reflective finish, in comparison to ruthenium's gun-metal grey. It is applied to a base of platinum or gold to give a shiny white finish, although the coating is so thin that it wears away over time. Being rarer than gold or platinum, rhodium is sometimes used to represent the very highest of honours; in 1979, Sir Paul McCartney was awarded a rhodium-plated disc to celebrate his achievement in becoming history's all-time bestselling recording artist and songwriter, with the exclusive metal signalling the extreme height of his success.

Rhodium plays a key role in reducing pollution from cars, in the catalytic converters that transform toxic exhaust gases into carbon dioxide and water vapour.

# Palladium

DISCOVERY DATE: 1802     DISCOVERED BY: William Hyde Wollaston

Palladium is a shiny, silver-coloured metal that does not tarnish when exposed to the air. It is the least dense member of the platinum group of metals and has the lowest melting point of them. It was discovered by an English chemist, William Hyde Wollaston in 1802 and named after the asteroid Pallas, which had been observed by Heinrich Olbers, a German astronomer, in 1802.

Like its next-door neighbour, rhodium, the main use of palladium is in catalytic converters for vehicles, where it acts as a catalyst to transform harmful gases into the non-toxic substances nitrogen, carbon dioxide and water vapour. Other applications for palladium are in jewellery, where it is alloyed with gold to form white gold, and in dental fillings, surgical instruments, spark plugs for aircraft and ceramic capacitors for electronic devices.

Palladium is a soft, ductile member of the platinum group of metals in the periodic table.

| 46 | |
|----|--|
| **Pd** | |
| Palladium | |
| 106.42 | |

| | |
|---|---|
| **Atomic number:** | 46 |
| **Group:** | Group 10 |
| **Period:** | Period 5 |
| **Block:** | d-block |
| **Atomic mass:** | 106.42 u |

| | |
|---|---|
| **Melting point:** | 1554.9 °C, 2830.82 °F |
| **Boiling point:** | 2963 °C, 5365 °F |
| **Density:** | 12.023 g/cm³ (near room temperature) |
| **Appearance:** | Silvery-white metal |

Palladium can be alloyed with gold to produce attractive 'white gold' jewellery.

One unusual feature of palladium is its ability to store vast amounts of hydrogen, through adsorption. Palladium can hold up to 900 times its own volume of hydrogen, which means that in principle it would be a very effective tool for storing hydrogen for other purposes. However, the rarity and expense of palladium means that this solution is not practical on a large scale.

Palladium(II) chloride ($PdCl_2$) plays a vital role in homes and workplaces thanks to its presence in carbon monoxide detectors. Because it catalyzes carbon monoxide gas – which is deadly but has no noticeable odour or colour – into carbon dioxide, it can be used to trigger an alarm in the event of a CO leak.

# Silver

DISCOVERY DATE: Approximately 3000 BCE

While there are multiple elements named after countries (such as americium, francium and polonium), silver is unique in having given its name to a country. When the people of Argentina declared independence from Spain in the early nineteenth century, they formally adopted the name of the metal their nation was so rich in for their new republic.

The chemical symbol of silver, Ag, comes from its Latin name, argentum, which echoes across Romance languages today with French *argent* and Italian *argento*. Its English name has roots in the Old English word 'siolfor' or 'seolfor'.

Silver has been prized for its shiny white reflective surface since prehistoric times, and has been used to make jewellery and other decorative objects for millennia. As well as being beautiful, silver is more electrically conductive than any other element,

Pure silver is rarely found in nature; usually silver occurs in combination with other metals or in a variety of minerals.

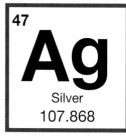

| | |
|---|---|
| **Atomic number:** | 47 |
| **Group:** | Group 11 |
| **Period:** | Period 5 |
| **Block:** | d-block |
| **Atomic mass:** | 107.868 u |
| **Melting point:** | 961.78 °C, 1763.2 °F |
| **Boiling point:** | 2162 °C, 3924 °F |
| **Density:** | 10.49 g/cm³ (near room temperature) |
| **Appearance:** | Shiny silvery-white metal |

**47**

**Ag**

Silver

107.868

and more thermally conductive than any other metal. It is also the best-known reflector of visible light, and silver-plated glass is used in specialist mirrors. As silver is relatively soft, it is usually alloyed with another metal for uses such as trophies, ornaments and jewellery. The best-known alloy is sterling silver, which combines 92.5 per cent silver with 7.5 per cent of a stronger metal, usually copper.

Because of their sensitivity to light, silver bromide and iodide are used in film photography, and this same characteristic is what makes photochromic spectacles react to changing light conditions: the lenses are embedded with silver chloride and halide, which change shape and become darker when exposed to sunlight, changing back again when the wearer's surroundings become less dazzling.

Silver has a stunning reflective finish and can be forged and fused into intricate shapes. As a result, it has been valued by jewellery makers for thousands of years.

# Cadmium

DISCOVERY DATE: 1817    DISCOVERED BY: Friedrich Stromeyer

Cadmium is a silver-coloured metal with a bluish tinge. It is ductile and corrosion-resistant, and shares many properties with the element directly above it in the periodic table, zinc. Its name comes from the Latin word cadmia, meaning 'calamine', a mineral that contains cadmium.

The primary use of cadmium today is in batteries, especially in rechargeable nickel-cadmium batteries; this application accounted for 86 per cent of cadmium use in 2009. Other functions of cadmium are in electroplating, to protect steel components from corrosion, in the control rods of nuclear reactors and, more recently, in the screens of QLED (quantum dot light-emitting diode) televisions.

In the past cadmium was also widely used as a paint pigment, producing a range of vibrant and long-lasting yellows, oranges and reds. The French Impressionist artist

**Cadmium shares some of the characteristics of zinc and mercury and is present in most zinc ores.**

| 48 | |
|---|---|
| **Cd** | |
| Cadmium | |
| 112.414 | |

| | |
|---|---|
| Atomic number: | 48 |
| Group: | Group 12 |
| Period: | Period 5 |
| Block: | d-block |
| Atomic mass: | 112.414 u |

| | |
|---|---|
| Melting point: | 321.07 °C, 609.93 °F |
| Boiling point: | 767 °C, 1413 °F |
| Density: | 8.65 g/cm³ (near room temperature) |
| Appearance: | Silvery blue-grey metal |

The first nickel–cadmium batteries were invented in 1899 and batteries with the same ingredients are still in use today. However, they are being phased out in some territories due to cadmium's toxicity.

Claude Monet, for example, employed cadmium yellow paints in his artworks to exquisite effect.

Despite its many useful properties, cadmium is being phased out of many processes due to its high toxicity. In the human body, exposure to cadmium is associated with kidney disease, high blood pressure and damage to the respiratory tract, and it may also increase the risk of cancer and osteoporosis. One particular problem with cadmium is that it accumulates in our bodies over time, so even a minimal regular intake can cause difficulties later in life. It also accumulates in many plants, including rice, cabbage, lettuce and tobacco. Because of this, it is vital that objects containing cadmium are disposed of safely, to stop this dangerous element being absorbed into the soil and, from there, into the food chain.

# Indium

DISCOVERY DATE: 1863   DISCOVERED BY: Ferdinand Reich and Hieronymous Richter

Although it is a silvery-grey metal, indium is named after the colour indigo. This is because of the bright indigo-coloured line that this metal revealed in its spectrum to the German researchers who first discovered it, Ferdinand Reich and Hieronymous Richter. Reich was colour-blind and had asked Richter to help him with the identification of this potential new element. Four years later, the friendship between the two scientists cooled when Reich learned that Richter had claimed himself to be the sole discoverer of indium at an exhibition in Paris.

Indium is the 68th most abundant element in the Earth's crust, and is usually found as a trace constituent of well-known ore minerals rather than in a pure form or in its own minerals. Most commercially produced indium comes as a by-product of zinc refining, and the chief countries producing indium are China, South Korea, Japan and Canada.

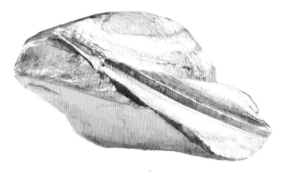

Indium, one of the softest elements, emits a distinctive 'crying' sound when it is bent.

| 49 |
| --- |
| **In** |
| Indium |
| 114.82 |

| Atomic number: | 49 |
| --- | --- |
| Group: | Group 13 |
| Period: | Period 5 |
| Block: | p-block |
| Atomic mass: | 114.82 u |

| Melting point: | 156.60 °C, 313.88 °F |
| --- | --- |
| Boiling point: | 2072 °C, 3762 °F |
| Density: | 7.31 g/cm³ (near room temperature) |
| Appearance: | Silvery-grey metal |

Indium has been designated a technology-critical element because it is used in so many electronic devices. The compound indium-tin oxide, known as ITO, sticks to glass and conducts electricity, and these properties make it an essential asset in the manufacture of LCDs (liquid crystal displays) and touchscreens, such as those on our mobile phones. It is also used in alloys, semiconductors and other electrical components.

Unusually, indium makes a 'crying' or creaking noise when it is bent. This is caused by the crystalline structure disintegrating and reforming as the metal changes shape. Indium's next-door neighbour in the periodic table, tin, has the same property.

As a semiconductor, indium has a wide range of applications. These thin-film solar cells use the photovoltaic compound copper indium gallium selenide (CIGS).

# Tin

DISCOVERY DATE: Approximately 2100 BCE

Tin is a silvery-coloured metal that is soft enough to be cut or bent easily. On bending, tin makes a 'crying' sound, like its neighbour at number 49 on the periodic table, indium. Its symbol, Sn, comes from *stannum*, the Latin word for 'tin' (although until the fourth century CE, this word signified an alloy of silver and lead).

Tin is sometimes called a 'poor' metal due to its softness, but when alloyed with other metals, it has been valued for its combination of strength and malleability for thousands of years. When tin is combined with copper, at a ratio of 12.5 per cent tin and 87.5 per cent copper, bronze is created – a strong, shiny metal whose uses range from weaponry to statuary and beyond. Another widespread and useful tin alloy is pewter, which is formed from 85–90 per cent tin in combination with antimony, copper, bismuth and sometimes silver.

Tin, a malleable silvery-white metal, was combined with copper to bring about the innovations of the Bronze Age, from *c.* 3300 BCE to 1200 BCE.

| | |
|---|---|
| **Atomic number:** | 50 |
| **Group:** | Group 14 |
| **Period:** | Period 5 |
| **Block:** | p-block |
| **Atomic mass:** | 118.710 u |
| **Melting point:** | 231.93 °C, 449.47 °F |
| **Boiling point:** | 2602 °C, 4716 °F |
| **Density:** | 7.265 g/cm³ (white, beta, near room temperature), 5.769 g/cm³ (grey, alpha, near room temperature) |
| **Appearance:** | Silvery-white metal (beta); grey metal (alpha) |

**50**

# Sn

Tin

118.710

There are also many modern applications of tin-based alloys. Superconducting magnets are manufactured from an alloy of niobium and tin, soft solders are made from tin alloyed with lead and indium tin oxide is used to create conductive coatings for the glass screens of electronic gadgets.

Another important property of tin is that it resists corrosion, and because of this, it has long been used as a plating for other metals, to protect them from damage by air or water. This resistance to corrosion, combined with its low toxicity to humans, has made tin a useful coating for the inside and outside of 'tin cans' – which are primarily made of steel, with a tin coating. Tin is also often used to plate the interior surfaces of copper pans, to prevent harmful reactions between the copper and acidic foods that might attack it.

From the mid-nineteenth century, colourful tinplated toys brought joy to children before plastic toys gradually took their place a hundred years later.

# Antimony

DISCOVERY DATE: Approximately 1600 BCE

Antimony, like arsenic above it and tellurium to the right of it in the periodic table, is a metalloid element, meaning that it shares some properties with metals and some with non-metals. It is not usually found in its elemental form in nature, and one interpretation of its name reflects this, coming from the Greek *anti-monos*, meaning 'against aloneness'. Another slightly more dramatic theory is that the name springs from *anti-monachos*, implying that the element antimony kills monks. The logic behind this is that early alchemists were monks, and some of them may have been poisoned by antimony due to its toxicity – but this feels a little fantastical, and its tendency to form alloys seems to have more scientific justification.

Antimony is often found in compounds involving sulfur, and its chemical symbol, Sb, comes from the Latin word *stibium*, or 'stibnite' in English ($Sb_2S_3$), the

Antimony is less well known than arsenic, which sits directly above it in the periodic table, but it is just as poisonous.

| | |
|---|---|
| **51** | |
| **Sb** | |
| Antimony | |
| 121.760 | |

| | |
|---|---|
| Atomic number: | 51 |
| Group: | Group 15 |
| Period: | Period 5 |
| Block: | p-block |
| Atomic mass: | 121.760 u |
| Melting point: | 630.63 °C, 1167.13 °F |
| Boiling point: | 1635 °C, 2975 °F |
| Density: | 6.697 g/cm³ (near room temperature) |
| Appearance: | Silvery-grey semi-metal |

**Sb**

There are many theories about the death of Mozart, as depicted here by Charles Edward Chambers. One is that he dosed himself with patent medicine containing antimony, without realising that it was dangerously poisonous.

naturally occurring mineral in which antimony is most commonly found. The main applications of antimony in industry today are in alloys with lead and tin, where it adds hardness and strength, in the manufacture of semiconductors, and as an additive to flame retardants.

Like arsenic, antimony is poisonous, and in humans it causes symptoms similar to arsenic poisoning, including headaches, dizziness, depression, vomiting and damage to the liver and kidneys. It has featured in many murder cases, and also in the famous unsolved case of the death of eminent British lawyer Charles Bravo in 1876. Numerous theories thrived as to how he had been poisoned, one of which was that he was gradually administering antimony to his wife in order to kill her, but then accidentally took a fatal dose himself from the wrong bottle in his medicine cabinet.

# Tellurium

DISCOVERY DATE: c. 1783    DISCOVERED BY: Franz-Joseph Müller von Reichenstein

In October 2023, the James Webb Space Telescope observed a kilonova, or the collision of two neutron stars, which unleashed a burst of gamma rays so bright that it was a million times more dazzling than the Milky Way. Astronomers were able to analyze the infrared signatures of the elements created in this vast explosion of energy, and they identified tellurium among them, along with other actinides and lanthanides, and the more commonly found iodine and thorium. This discovery is exciting because, while it was already well known that lighter elements come together through fusion in the cores of stars to create heavier elements, this was the first time that the birth of heavier elements through the collision of neutron stars had ever been observed.

Tellurium is not needed by the human body, and most of the 600 micrograms of it that we consume on an average day is excreted through urine or passed through the

Tellurium takes its name from the Latin *tellus*, meaning 'earth', but it is actually far more abundant in the Universe as a whole than in our planet's crust.

| 52 | |
|---|---|
| **Te** | |
| Tellurium | |
| 127.60 | |

| | |
|---|---|
| **Atomic number:** | 52 |
| **Group:** | Group 16 |
| **Period:** | Period 5 |
| **Block:** | p-block |
| **Atomic mass:** | 127.60 u |

| | |
|---|---|
| **Melting point:** | 449.51 °C, 841.12 °F |
| **Boiling point:** | 988 °C, 1810 °F |
| **Density:** | 6.24 g/cm³ (near room temperature) |
| **Appearance:** | Grey powder |

gut. Although it is not toxic, it does have the unpleasant side effect of causing bad breath, with a garlic-like smell. In 1884, volunteers in a study who had taken 15mg (0.000529oz) doses of the tellurium oxide were found to still have detectable 'tellurium breath' as much as eight months later. Taking extra doses of vitamin C is thought to help with this problem, but the safest course of action would be to try to avoid exposure in the first place.

Tellurium can be used in fibre optic cables, where it increases the transmission rate.

# Iodine

Iodine is the 61st most abundant element on Earth, and the heaviest stable member of Group 17, the halogens. It was discovered in 1811 by French chemist Bernard Courtois, who, after extracting potassium chloride crystals from seaweed, added sulfuric acid to the remaining liquid, and saw a purple gas emerge, which condensed into shiny black crystals on cooling. Two years later, Courtois's discovery was confirmed to be a new element, and named after the Greek word *iodes*, meaning 'violet', because of its vibrantly coloured vapour.

Iodine is an essential element for animals because of the role it plays in the hormones triiodiothyronine and thyroxine (known as T3 and T4), which regulate the genes that control metabolism, growth and development. Many people are able to take in sufficient iodine for their biological needs by eating dairy products, eggs, fish and shellfish. People

Iodine crystals look black, but as a gas, iodine has the characteristic purple colour that gives it its name.

| 53 | |
|---|---|
| **I** | |
| Iodine | |
| 126.904 | |

| | |
|---|---|
| Atomic number: | 53 |
| Group: | Group 17 |
| Period: | Period 5 |
| Block: | p-block |
| Atomic mass: | 126.904 u |
| Melting point: | 113.7 °C, 236.66 °F |
| Boiling point: | 184.3 °C, 363.7 °F |
| Density: | 4.933 g/cm$^3$ (near room temperature) |
| Appearance: | Black shiny crystalline solid, purple gas |

All animals need iodine in their diets; a glass of milk is an ideal source, along with other dairy products, eggs, fish and seaweed.

with a vegan diet may require iodine supplements, as although iodine is present in cereals and grains, the amount of the element contained varies depending on the soil that the crops were grown in.

Iodine deficiency affects millions of people around the world, particularly in developing countries and inland regions with limited access to foods from the sea. This causes a range of issues including weight gain, fatigue, depression and goitre (a swollen neck due to an enlarged thyroid gland), as well as intellectual disabilities in babies and young children.

Commercially, iodine is used as a catalyst, especially in the production of acetic acid, as an antiseptic, in dyes, printing inks and photographic chemicals, and in animal feed. The largest producers of iodine are Japan and Chile, using a process of heating, purifying and acidifying brine.

Noble gas

# Xenon

DISCOVERY DATE: 1898    DISCOVERED BY: William Ramsay and Morris Travers

Like the other members of Group 18, xenon is a colourless, odourless, unreactive gas. It was named after the Greek word *xenos*, meaning 'stranger', due to the fact that it had proved hard to find. Only trace amounts of xenon are present in the Earth's atmosphere, making it an expensive resource.

Although it is a member of the noble gas group, xenon is not totally inert. In 1962, Neil Bartlett and his team at the University of British Columbia in Canada, produced xenon hexafluoroplatinate, the first ever known compound of a noble gas, and to date, more than 100 different xenon-based compounds have been made. When electricity is passed through it in a gas-filled tube, xenon glows with a vivid blue colour. Its light-emitting properties have a wide range of applications, including in photographic flashbulbs, sunbed tubes and lamps that kill bacteria in food preparation contexts.

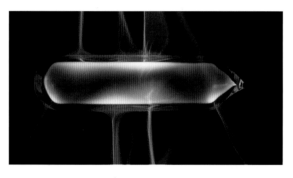

Xenon gas glows with a light blue radiance when excited in a discharge tube.

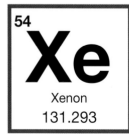

54

# Xe

Xenon

131.293

| Atomic number: | 54 |
|---|---|
| Group: | Group 18 |
| Period: | Period 5 |
| Block: | p-block |
| Atomic mass: | 131.293 u |

| Melting point: | −111.75 °C, −169.15 °F |
|---|---|
| Boiling point: | −108.099 °C, −162.578 °F |
| Density: | 0.005894 g/cm³ (near room temperature) |
| Appearance: | Colourless gas |

Xenon has no biological role and is non-toxic, although it is unusual in being able to pass through the blood-brain barrier. As a result of this, when breathed in with oxygen, it can cause mild to complete anaesthesia, and it has been used surgically for this purpose. Like helium, xenon changes the quality of your voice if you breathe it in – although with the opposite effect. While helium enhances high-frequency sounds produced by the vocal cords, xenon increases the resonance of low-frequency sounds, giving the speaker a deeper-sounding voice than normal.

**Xenon car headlights are brighter than their halogen equivalents and can last for ten years without needing to be replaced.**

# Caesium

DISCOVERY DATE: 1860     DISCOVERED BY: Gustav Kirchhoff and Robert Bunsen

Caesium is a soft, gold-coloured metal that is solid at standard room temperate but will melt into a liquid on a moderately hot day, thanks to its low melting point of only 28.5°C (83.3°F). Like all members of Group 1 of the periodic table, it is extremely reactive. It can catch fire spontaneously in air, and explodes violently when it makes contact with water.

Caesium's name comes from caesius, a Latin word meaning 'bluish grey', due to the bright blue lines observed in its emossion spectrum by its German discoverers, Gustav Kirchhoff and Robert Bunsen. This element's name offers further opportunity for transatlantic discussion, as the International Union of Pure and Applied Chemistry recommends the spelling 'caesium', however, the American Chemical Society has preferred the more concise 'cesium' since 1921.

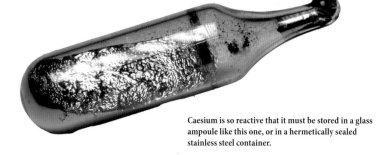

Caesium is so reactive that it must be stored in a glass ampoule like this one, or in a hermetically sealed stainless steel container.

| 55 | |
|---|---|
| **Cs** | |
| Caesium | |
| 132.905 | |

| | |
|---|---|
| **Atomic number:** | 55 |
| **Group:** | Group 1 |
| **Period:** | Period 6 |
| **Block:** | s-block |
| **Atomic mass:** | 132.905 u |

| | |
|---|---|
| **Melting point:** | 28.5 °C, 83.3 °F |
| **Boiling point:** | 671 °C, 1240 °F |
| **Density:** | 1.93 g/cm³ (near room temperature) |
| **Appearance:** | Gold-coloured metal |

The CS2 atomic clock CS2 at the German National Metrology Institute, Germany differs by just one second from the ideal time in about 2.5 million years.

There is no known biological role for caesium. One of its radioactive isotopes, caesium-137, has been used as a cancer treatment, but this same isotope has also been ingested involuntarily by large populations following nuclear tests and accidents, such as the Chernobyl disaster in 1986 and the Fukushima Daiichi disaster in 2011.

One of the best-known uses of caesium is in atomic clocks, which provide a shared time reference for the internet, GPS services and mobile phones. These clocks are the most accurate measure of time that have been created, and the very latest models have an accuracy level of one second in 20 million years. Caesium is also used in the production of special optical glass, as a 'getter' in vacuum tubes, and as a catalyst in the hydrogenation of organic compounds.

# Barium

DISCOVERY DATE: 1808    DISCOVERED BY: Humphry Davy

Barium is a soft, silver metal that oxidizes rapidly in air, giving a dark grey outer layer. It is named after the ore baryte, from which it was first isolated in 1808 by the English chemist Humphry Davy. Baryte's name itself comes from the Greek word *barys*, meaning 'heavy'. Baryte is still the main commercial source of barium today, along with witherite, or barium carbonate, and the main deposits of these minerals are found in the UK, Romania and the former Soviet Union.

Barium has only a small range of industrial applications. It can be used as a 'getter' to remove unwanted gases from vacuum tubes, although since the gradual replacement of cathode-ray tube televisions by flatscreen models, the demand for this function has declined. It is also used in high-temperature superconductors, and in an alloy with nickel for sparkplug wiring. More festively, barium salts give their

Barium sulfate crystal, also known as barite, was the substance that made up the famous 'Bologna stones' of the early 1600s, which glowed in the dark after being exposed to sunlight during the day.

| 56 | |
|---|---|
| **Ba** | |
| Barium | |
| 137.33 | |

| | |
|---|---|
| **Atomic number:** | 56 |
| **Group:** | Group 2 |
| **Period:** | Period 6 |
| **Block:** | s-block |
| **Atomic mass:** | 137.33 u |

| | |
|---|---|
| **Melting point:** | 727 °C, 1341 °F |
| **Boiling point:** | 1845 °C, 3353 °F |
| **Density:** | 3.51 g/cm³ (near room temperature) |
| **Appearance:** | Silvery-grey metal |

colours to firework explosions: barium nitrate produces a yellowy green and barium monochloride creates a brilliant green.

Another well-known application of barium is in the form of barium sulfate, which is used in so-called 'barium meals' for medical scans of the gastrointestinal tract. When swallowed or given as an enema, a barium sulfate suspension absorbs X-rays, which pass directly through all other parts of the body. The images produced can be used to diagnose diseases and abnormalities of the digestive system.

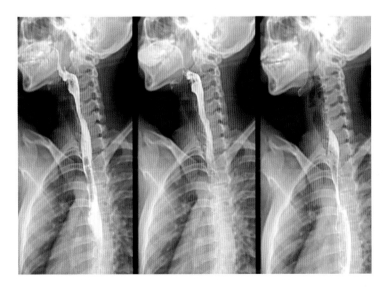

A 'barium swallow' procedure allows radiographers to see a patient's oesophagus and stomach, which are cannot normally be seen in standard X-ray images.

# Lanthanum

DISCOVERY DATE: 1839    DISCOVERED BY: Carl Gustaf Mosander

So far, our journey through the periodic table has proceeded in an orderly fashion, moving step by step across the groups and down the periods. However, with element 57, lanthanum, we have to leap out of the main table and into its annexe along the bottom: the lanthanide group (with the actinide group lining up neatly below it). The elements in this group used to be called the 'rare earth elements', but many of them are not rare at all: as one example, lanthanum is the 28th most abundant element in the Earth's crust

Lanthanum, a soft, silvery-white reactive metal, was discovered by the Swedish chemist Carl Gustaf Mosander in 1839. Its name comes from the Greek word *lanthanein*, meaning 'to lie hidden', because it had been found in cerium ore, and it was so similar to cerium that it was hard to tell the two elements apart.

Lanthanum, the first of the lanthanide series, is a silvery-coloured metal that tarnishes slowly when exposed to air.

**57**

**La**

Lanthanum

138.905

| | |
|---|---|
| Atomic number: | 57 |
| Group: | Lanthanides |
| Period: | Period 6 |
| Block: | f-block |
| Atomic mass: | 138.905 u |

| | |
|---|---|
| Melting point: | 920 °C, 1688 °F |
| Boiling point: | 3464 °C, 6267 °F |
| Density: | 6.162 g/cm³ (near room temperature) |
| Appearance: | Silvery-white metal |

Bastnäsite, named after the Bastnäs mine in Västmanland, Sweden, is a key ore in the production of lanthanum.

In the late nineteenth century, lanthanum oxides were used in the mantles of gas lanterns, and later, when the film industry developed, lanthanum was employed to improve the quality of light emitted by the carbon arc lamps in studios. Today, a key application of this element is in nickel-metal hydride (NiMH) batteries used in consumer electronics and hybrid electric cars.

# Cerium

DISCOVERY DATE:1803     DISCOVERED BY: Jöns Jacob Berzelius and Wilhelm Hisinger

Cerium is the second element in the lanthanide series and the 25th most abundant element on Earth. It was discovered by Jöns Jacob Berzelius and Wilhelm Hisinger in Sweden in 1803, and independently by Martin Klaproth in Germany in the same year. Then in 1839, Sweden's Carl Gustaf Mosander isolated pure cerium metal for the first time, from its parent ore, cerite.

Cerium, which Berzelius named after the recently discovered asteroid Ceres, is a grey metal that is soft and ductile, with four naturally occurring isotopes and 26 radioactive ones. It is found in various minerals, particularly in bastnäsite and monazite, and commercial production is based on heating these ores and treating them with hydrochloric and sulfuric acids.

Cerium is the most abundant of the lanthanides, which are also known as the rare earth elements.

| 58 | |
|---|---|
| **Ce** | |
| Cerium | |
| 140.116 | |

| | |
|---|---|
| Atomic number: | 58 |
| Group: | Lanthanidea |
| Period: | Period 6 |
| Block: | f-block |
| Atomic mass: | 140.116 u |
| Melting point: | 795 °C, 1463 °F |
| Boiling point: | 3443 °C, 6229 °F |
| Density: | 6.770 g/cm³ (near room temperature) |
| Appearance: | Grey metal |

Cerium has a huge range of uses. It is found in catalytic converters, self-cleaning ovens and sunlight-stable clear polymers.

Mischmetal alloy, which contains a mix of lanthanide elements combined with about 5 per cent iron, is used as the 'flint' for many lighters, because the cerium in the alloy creates sparks when another metal is dragged across it. Cerium(III) oxide is used in catalytic converters and on the walls of self-cleaning ovens, where it prevents cooking residues from accumulating. And because of its bright red colour, cerium(III) sulfide is valued as a non-toxic safer alternative to older cadmium-based pigments, and it is also stable at high temperatures.

Cerium is not known to have any biological function in humans, although it and some of the other lanthanides are essential to certain bacteria that live in in volcanic mudpots. In humans, cerium nitrate is a useful treatment for third-degree burns, although it is toxic in large amounts.

# Praseodymium

DISCOVERY DATE: 1885    DISCOVERED BY: Carl Auer von Welsbach

Praseodymium, the third element in the lanthanide series, first came into the world as a twin, and would later become a twin again. If that sounds complicated, it's because the lanthanide elements are very similar, quite hard to tell apart, and even to separate from each other in some cases.

In 1841, Swedish chemist Carl Gustaf Mosander, who had previously isolated lanthanum and cerium, found what he believed to be a new element. He gave it the name didymium, from the Greek word for 'twin', because of its similarity to the first two lanthanide elements. However, in the following decades, scientists investigated this new substance more deeply, and in 1885, the Austrian scientist Carl Auer von Welsbach demonstrated that it was composed of two separate elements. One of these was named praseodymium, meaning 'green twin', because it oxidises with a green

Praseodymium, the 'green twin' of the periodic table, forms a green oxide coating when exposed to air.

| 59 | |
|---|---|
| **Pr** | |
| Praseodymium | |
| 140.908 | |

| | |
|---|---|
| Atomic number: | 59 |
| Group: | Lanthanides |
| Period: | Period 6 |
| Block: | f-block |
| Atomic mass: | 140.908 u |
| Melting point: | 935 °C, 1715 °F |
| Boiling point: | 3130 °C, 5666 °F |
| Density: | 6.77 g/cm³ (near room temperature) |
| Appearance: | Greyish-white metal |

coating when exposed to air, and the other was called neodymium, or 'new twin', to reflect its birth from the former didymium.

One major application of praseodymium is in alloys with magnesium, where its strength is useful for the construction of aircraft engines. It is also used in protective goggles, for the yellow tint it gives to the lenses, protecting the wearer's eyes from damage by light and infrared radiation during welding and glass-blowing. Like lanthanum, praseodymium is also used in carbon arc lights for studios and projectors.

**Praseodymium-tinted lenses help to protect the wearer's eyes from being damaged by the bright light from welding equipment.**

# Neodymium

DISCOVERY DATE: 1885     DISCOVERED BY: Carl Auer von Welsbach

Neodymium was discovered at the same time as its neighbour praseodymium, by the Austrian chemist Carl Auer von Welsbach, who used spectroscopic analysis to prove that the element known as didymium in fact consisted of two combined elements.

Like praseodymium, neodymium is used to give strength to magnesium alloys, and also to give its colour to protective goggles. While its 'twin', praseodymium, turns glass and ceramics yellow, neodymium compounds can create colours ranging from grey-blue to violet and pink. Glass made with neodymium oxide ($Nd_2O_3$) has a remarkable property, in that it appears to be lavender-coloured in daylight or incandescent light, but changes to a pale blue under fluorescent light.

Another important application of this element is in the production of neodymium magnets, made from an alloy of neodymium, iron and boron ($Nd_2Fe_{14}B$). These are the

Neodymium is a shiny silvery metal that tarnishes rapidly when exposed to air or moisture.

**60**

**Nd**

Neodymium

144.242

| | |
|---|---|
| **Atomic number:** | 60 |
| **Group:** | Lanthanides |
| **Period:** | Period 6 |
| **Block:** | f-block |
| **Atomic mass:** | 144.242 u |
| **Melting point:** | 1024 °C, 1875 °F |
| **Boiling point:** | 3074 °C, 5565 °F) |
| **Density:** | 7.01 g/cm³ (near room temperature) |
| **Appearance:** | Silvery-white metal |

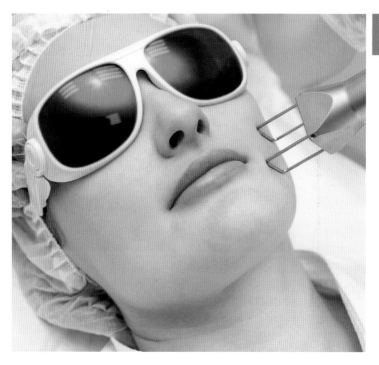

Neodymium-doped yttrium aluminium garnet (Nd:YAG) lasers are used in cosmetic treatments such as laser hair removal.

strongest known permanent magnets, and they are perfect for use in contexts where strong magnetic power is needed in combination with a small size and light weight, such as in headphones, computer hard disks and the pick-ups of electric guitars. They are also used in the motors of electric cars and some commercial wind turbines. These magnets are so strong that they can be highly dangerous: two palm-sized magnets can snap together with enough force to break human bones, so they need to be handled and stored with extreme care.

# Promethium

DISCOVERY DATE: 1942    DISCOVERED BY: Jacob A. Marinsky, Lawrence E. Glendenin & Charles D. Coryell

Most of the 'rare earth elements', also known as the lanthanide series, are in fact not rare at all, but promethium truly lives up to this label. At any given time, there are only 500–600g (17–21oz) of promethium naturally occurring on Earth, because it is radioactive and all its isotopes have short half-lives, with the result that it quickly decays out of existence.

In the natural world, promethium is created by the alpha decay of europium-151, and from the spontaneous fission of uranium. It was first produced in 1945 at the Oak Ridge National Laboratory in Tennessee by Jacob A. Marinsky, Lawrence E. Glendenin and Charles D. Coryell from the analysis and separation of fission products of uranium fuel, although they did not publicly announce their discovery until 1947. Then in 1963, researchers at the same laboratory succeeded in producing 10g (0.3oz) of promethium

The mineral uraninite is the host for most of Earth's promethium, containing about 200 picograms of promethium per kilo of uraninite.

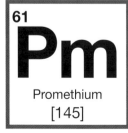

**61**

**Pm**

Promethium

[145]

| | |
|---|---|
| **Atomic number:** | 61 |
| **Group:** | Lanthanides |
| **Period:** | Period 6 |
| **Block:** | f-block |
| **Atomic mass:** | [145] |

| | |
|---|---|
| **Melting point:** | 1042 °C, 1908 °F |
| **Boiling point:** | 3000 °C, 5432 °F |
| **Density:** | 7.26 g/cm³ (near room temperature) |
| **Appearance:** | Silvery metal |

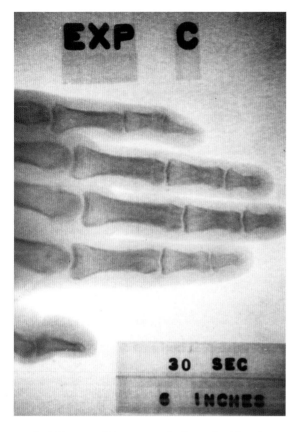

This image was produced in around 1963 by a portable X-ray system powered by promethium-147.

metal, which was a sufficient amount for them to be able to discover some of its key properties, including its melting point.

Due to its extreme rarity, promethium does not have many commercial applications, being largely confined to research purposes. However, it is used in specialized atomic batteries for pacemakers, radios and guided missiles. It was once applied on the dials of clocks and watches to make them glow in the dark, after radium had been ruled out due to its damaging radioactivity, but because of its short half-life, it didn't retain its luminous properties for long enough to be a successful alternative.

# Samarium

DISCOVERY DATE:1879     DISCOVERED BY: Paul-Émile Lecoq de Boisbaudran

Samarium is a silvery-white metal that oxidizes slowly in air. Its name comes from the mineral in which it was found, samarskite, and that mineral itself was named after Colonel Vassili Samarsky-Bykhovets, the chief of the Russian Corps of Mining Engineers, who had first given permission for German researchers Gustav and Heinrich Rose to study rock samples from the Urals. Various scientists were studying lanthanide-containing minerals in the second half of the nineteenth century, and a number announced their discovery of the missing element 62, but the official discoverer of samarium is French chemist Paul-Émile Lecoq de Boisbaudran, who also discovered gallium and dysprosium.

Like neodymium, samarium is used to make magnets that are both small and incredibly strong, for use in guitar pick-ups, headphones and motors. These magnets

Samarium, a shiny silver-white metal, is the 40th most abundant metal in the Earth's crust.

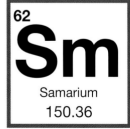

**62**
## Sm
Samarium
150.36

| | |
|---|---|
| **Atomic number:** | 62 |
| **Group:** | Lanthanides |
| **Period:** | Period 6 |
| **Block:** | f-block |
| **Atomic mass:** | 150.36 u |
| **Melting point:** | 1072 °C, 1962 °F |
| **Boiling point:** | 1900 °C, 3452 ° |
| **Density:** | 7.52 g/cm³ (near room temperature) |
| **Appearance:** | Silvery-white metal |

are made from a compound of samarium and cobalt, and they are 10,000 times stronger than an iron magnet of the same size. Although they are not quite as powerful as neodymium-iron-boron magnets, they are preferred in some contexts because they maintain their magnetism better at high temperatures.

Some isotopes of samarium have such a long half-life that they can be used to date meteorites and rocks. This is because, over time, isotopes of samarium decay into isotopes of neodymium, and the degree of change enables scientists to determine the age of a sample. In medicine, the isotope Samarium-153 is used in a drug with the trade name Quadramet to kill cancerous cells in cancers of the lung, prostate, breast and bones.

**Samarium magnets are used in headphones, small motors and for the pickups in electric guitars.**

# Europium

DISCOVERY DATE:1901     DISCOVERED BY: Eugène-Anatole Demarçay

Europium is a soft, ductile, silver-coloured metal that oxidizes in air to form a dark outer coating. It reacts vigorously with water in a similar way to calcium, to produce europium hydroxide and hydrogen gas. It was isolated for the first time by French chemist Eugène-Anatole Demarçay in 1901, although his compatriot Paul-Émile Lecoq de Boisbaudran had previously identified it in a samarium-gadolinium concentrate in 1892.

For commercial production, europium is extracted from lanthanide-rich ores such as bastnäsite and monazite. The main sources of these ores are at Bayan Obo in Inner Mongolia, China, and the Mountain Pass rare earth mine in California.

In comparison with other elements, europium only has a limited number of industrial applications. Two forms of europium oxide ($Eu_2O_3$)are used in the

Europium, named after the continent of Europe, is one of the rarest of the rare-earth metals.

| 63 | | |
|---|---|---|
| **Eu** | | |
| Europium | | |
| 151.964 | | |

| Atomic number: | 63 |
|---|---|
| Group: | Lanthanides |
| Period: | Period 6 |
| Block: | f-block |
| Atomic mass: | 151.964 u |

| Melting point: | 826 °C, 1519 °F |
|---|---|
| Boiling point: | 1529 °C, 2784 °F |
| Density: | 5.244 g/cm³ (near room temperature) |
| Appearance: | Silvery-white metal |

screens of our computers and flatscreen televisions: trivalent europium produces a red radiance, whereas bivalent europium creates a blue radiance. The red aspect of europium oxide is also used to create a warmer, more natural-looking glow in energy-saving light bulbs and modern street lamps, which can otherwise cast a very cold white light.

Europium compounds are also used as a security measure in banknotes, and especially appropriately, in euro notes. When ultraviolet light is shone onto the notes, electrons in the compounds become excited and fluoresce visibly. If a note was tested under UV light and did not reveal these hidden marks, it would clearly be a counterfeit.

This 50-euro banknote uses europium in its anti-counterfeiting measures.

# Gadolinium

DISCOVERY DATE:1880    DISCOVERED BY: Jean Charles Galissard de Marignac

Gadolinium is a malleable and ductile silvery-white metal in the lanthanide group of the periodic table. It is the sixth most abundant lanthanide element and the 41st most abundant element in the Earth's crust, discovered in 1880 by the Swiss chemist Jean Charles Galissard de Marignac, using spectroscopy. Its name comes from the mineral gadolinite, which itself was named after a Finnish chemist, Johan Gadolin.

One key application of gadolinium is in nuclear reactors, because it has the greatest capacity to capture neutrons of any known element. As a result, it is used in reactor control rods, and also as an emergency shut-down measure. Another somewhat surprising application of this element is in magnetic refrigerators. Unlike our domestic fridges, these refrigerators work on the basis of a gadolinium alloy heating up when a magnetic field is applied to it, and then cooling down – to a temperature *lower*

Gadolinium, a silvery-white metal in the lanthanide series, is commercially produced in China, the USA, Brazil, Sri Lanka, India and Australia.

| 64 | |
|---|---|
| **Gd** | |
| Gadolinium | |
| 157.25 | |

| | |
|---|---|
| Atomic number: | 64 |
| Group: | Lanthanides |
| Period: | Period 6 |
| Block: | f-block |
| Atomic mass: | 157.25 u |

| | |
|---|---|
| Melting point: | 1312 °C, 2394 °F |
| Boiling point: | 3273 °C, 5923 °F |
| Density: | 7.90 g/cm³ (near room temperature) |
| Appearance: | Silvery-white metal |

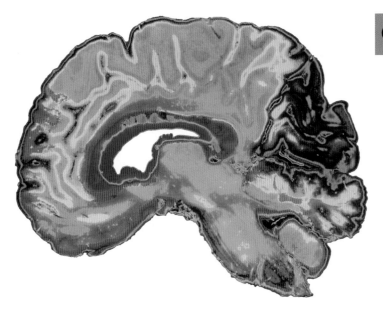

Gadolinium-based contrast agents are used to highlight brain tumours and other brain disorders in MRI scanning procedures.

than the original starting point – when the magnetic field is removed. This method of refrigeration is not yet available commercially, but its development could bring significant environmental and efficiency benefits in the future.

Gadolinium plays no known role in the human body, but it does have various medical applications. It is used in MRI scans as a contrast agent to make the images easier to see. Although pure gadolinium is toxic to humans, it is made safe for use as a drug through a process of chelation: this means that it is mixed with other chemical ions that neutralize its damaging effects while maintaining its ability to be seen in body tissue.

# Terbium

DISCOVERY DATE:1843     DISCOVERED BY: Carl Gustaf Mosander

Like ytterbium, yttrium and erbium, terbium is named after the Swedish town of Ytterby, close to the mine where the mineral containing it was found. Ytterby holds the record as the place with the greatest number of elements named after it – most other locations are lucky to have just one.

Terbium was discovered by Carl Gustaf Mosander in 1843, through analysis of ytterbium oxide, or 'yttria'. It is a silvery metal that is malleable and ductile, and soft enough to be cut with a knife. Terbium is one of the rarest of the lanthanides and is never found in a pure form in nature due to its reactivity. Instead, it is found in ores such as monazite, bastnäsite and euxenite.

Like europium, terbium is used in compounds applied to euro banknotes as an anti-counterfeiting measure, as it fluoresces under ultraviolet light. It is also used to produce

Terbium, the ninth member of the lanthanide group, never occurs in its pure form in nature and has to be extracted from ores such as gadolinite and monazite.

| 65 | |
|---|---|
| **Tb** | |
| Terbium | |
| 158.925 | |

| | |
|---|---|
| Atomic number: | 65 |
| Group: | Lanthanides |
| Period: | Period 6 |
| Block: | f-block |
| Atomic mass: | 158.925 u |

| | |
|---|---|
| Melting point: | 1356 °C, 2473 °F |
| Boiling point: | 3123 °C, 5653 °F |
| Density: | 8.23 g/cm³ (near room temperature) |
| Appearance: | Silvery-white metal |

Carl Gustaf Mosander (1797–1859) was a Swedish chemist who discovered three chemical elements: lanthanum, erbium and terbium.

green light in LED screens in the form of terbium-based phosphors. In combination with the red and blue lights provided by trivalent and bivalent europium-based compounds, terbium can produce a bright white light in trichromatic lamps, which for a given amount of energy are much more brilliant than incandescent light sources.

An alloy of terbium, iron and dysprosium with the trade name Terfenol-D has an unusual property called magnetostriction, which means it vibrates when an electrical current runs through it. As a result, it can transform any flat surface into a loudspeaker, an effect that it already used in actuators and naval sonar systems, and it may have further applications in the future.

# Dysprosium

DISCOVERY DATE:1886    DISCOVERED BY: Paul-Émile Lecoq de Boisbaudran

Dysprosium's name comes from the Greek word *dysprositos*, meaning 'hard to obtain', because French chemist Paul-Émile Lecoq de Boisbaudran needed more than 30 attempts to separate it from its oxide in 1886. In fact, it took until the 1950s before another chemist, Iowa University's Frank Spedding, was able to obtain relatively pure samples of erbium metal and its oxide. Pure dysprosium is never found in nature, but it is present in many minerals, such as xenotime, fergusonite, gadolinite, monazite and bastnäsite. Commercially, most dysprosium comes from monazite sand, as a by-product from the production of yttrium. Around 100 tonnes (110 tons) of dysprosium are produced annually of which 99 per cent comes from China.

The largest application of dysprosium, accounting for some 98 per cent of its use, is in magnets. When dysprosium is added to neodymium–iron–boron magnets,

Dysprosium, element 66 in the periodic table, was difficult to isolate but has proved to be highly valued in a range of commercial applications.

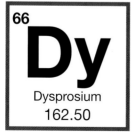

**66**
# Dy
Dysprosium
162.50

| Atomic number: | 66 |
| --- | --- |
| Group: | Lanthanides |
| Period: | Period 6 |
| Block: | f-block |
| Atomic mass: | 162.50 u |

| Melting point: | 1407 °C, 2565 °F |
| --- | --- |
| Boiling point: | 2562 °C, 4653 °F |
| Density: | 8.54 g/cm³ (near room temperature) |
| Appearance: | Silvery-white metal |

Neodymium–iron–boron magnets with added dysprosium are used in wind turbine generators.

replacing up to 6 per cent of the neodymium, it increases the magnet's coercivity, meaning that it can withstand an external magnetic field for longer without becoming demagnetized. These magnets are used in wind turbines and electric vehicles, with around 100g (3.5oz) of dysprosium needed for the motor of every electric car. In the longer term, as electric cars become more widespread, the supply of dysprosium will be too limited to keep up with the demand.

Another important application of dysprosium is in radiation dosimeters. The metal is added to crystals of calcium sulfate or calcium fluoride, which then fluoresce in the presence of ionizing radiation. The dosimeter measures the level of light emitted to indicate how much radiation is present.

# Holmium

DISCOVERY DATE: 1878 DISCOVERED BY: Per Teodor Cleve (Sweden), Marc Delafontaine & Louis Soret (Switzerland)

Like most of the other lanthanide elements, holmium is a silvery-white metal that is also ductile and fairly soft. One of its co-discoverers, Swedish chemist Per Teodor Cleve, was from Stockholm, and he named the new element after his home town, based on its Latin name of *Holmia*.

Of all the naturally occurring elements, holmium has the highest magnetic moment, although unlike iron, cobalt and nickel, it requires the presence of a magnetic field before it becomes magnetic itself. It is used in the most powerful magnets as a pole piece, and it is also part of the manufacturing process for some permanent magnets. Holmium's unusual magnetic properties have also been investigated for their potential as a data storage solution. In 2017, researchers from IBM created the world's smallest magnet, from just one single atom of holmium, and they were able to switch the poles

Holmium is the eleventh member of the lanthanide series and possesses a number of unusual magnetic properties.

| 67 | | |
|---|---|---|
| **Ho** | | |
| Holmium | | |
| 164.930 | | |

| Atomic number: | 67 |
|---|---|
| Group: | Lanthanides |
| Period: | Period 6 |
| Block: | l-block |
| Atomic mass: | 164.930 u |

| Melting point: | 1461 °C, 2662 °F |
|---|---|
| Boiling point: | 2600 °C, 4712 °F |
| Density: | 8.79 g/cm³ (near room temperature) |
| Appearance: | Soft silvery metal |

of its magnetism by passing an electrical current through it. This kind of technology, if developed successfully, could one day see several hard drives' worth of data stored on a device as small as a credit card.

Holmium is not essential for human life, and the average person consumes only about 1mg (0.00003527) of holmium per year. However, it does have a medical application as in combination with YAG (yttrium aluminium garnet) surgical lasers, which are used for operations such as the removal of kidney stones or small cancerous tumours. The light emitted by these lasers does not damage human eyesight, and this means that they are also suitable for use in eye surgery.

Holmium oxide is used to give a red colour to cubic zirconia, to create these garnet-like jewels.

# Erbium

DISCOVERY DATE:1842    DISCOVERED BY: Carl Gustaf Mosander

Erbium takes its name from the Swedish town of Ytterby, in honour of the mine where its ore was first sourced. Like so many other lanthanides, erbium is a soft and malleable silvery-coloured metal. It was discovered in 1842 by Carl Gustaf Mosander, who also first identified lanthanum and terbium.

Erbium salts fluoresce with a pink colour in natural ultraviolet light, for this reason, they are often used to colour camera filters, sunglass lenses and protective visors. This same property of erbium is also applied to give colour to jewels and porcelain, although it was only in the 1990s that erbium oxide became available at a cheap enough price to make this application commercially viable.

Another important use of erbium is in fibre-optic cables, where it amplifies the signals being transmitted and prevents the light in the signal from fading before

Like the other lanthanides, erbium is a silvery-white metal. It has a range of uses including fibre-optic cables, medical lasers and as a glass colourant.

| 68 | Er | Erbium | 167.259 |
| --- | --- | --- | --- |

| | |
| --- | --- |
| **Atomic number:** | 68 |
| **Group:** | Lanthanides |
| **Period:** | Period 6 |
| **Block:** | l-block |
| **Atomic mass:** | 167.259 u |

| | |
| --- | --- |
| **Melting point:** | 1529 °C, 2784 °F |
| **Boiling point:** | 2868 °C, 5194 °F |
| **Density:** | 9.066 g/cm³ (near room temperature) |
| **Appearance:** | Soft silvery metal |

Erbium oxide is used as a dopant to produce glass lenses that protect wearers from damaging their eyes.

it reaches its destination. When these cables are laid along the ocean floor, as well as transmitting vital and not-so-vital pieces of data from one landmass to another, they also attract the attention of sharks, who possess sensory organs that can detect electromagnetic fields. To avoid any shark-related data outages, these cables are wrapped in a Kevlar coating to protect them from any potential chewing.

Erbium is also used in medical lasers for cosmetic procedures and dentistry, and in the neutron-absorbing control rods in nuclear reactors.

# Thulium

DISCOVERY DATE:1879     DISCOVERED BY: Per Teodor Cleve

Thulium, a soft, silvery-grey metal, is the 13th member of the lanthanides series, and was discovered by Swedish chemist Per Teodor Cleve in 1879. It is the second least abundant of the lanthanides after promethium, and therefore truly deserves this group's alternative name 'rare earth elements'. As a result of this scarcity, it is very valuable. However, thulium has no one property or application that distinguishes it from the other lanthanides, and so there is not a huge demand for it.

Thulium's name comes from Thule, a fabled northern land in Greek mythology that no explorer had ever reached, and a name that is now associated with the Nordic countries. After its discovery in 1879 by Per Teodor Cleve in Sweden, it took more than 30 years for another scientist, American chemist Charles James, to isolate the

Thulium is a fairly soft lanthanide metal that tarnishes slowly when exposed to air.

| 69 | |
|---|---|
| **Tm** | |
| Thulium | |
| 168.934 | |

| | |
|---|---|
| **Atomic number:** | 69 |
| **Group:** | Lanthanides |
| **Period:** | Period 6 |
| **Block:** | f-block |
| **Atomic mass:** | 168.934 u |

| | |
|---|---|
| **Melting point:** | 1545 °C, 2813 °F |
| **Boiling point:** | 1950 °C, 3542 °F |
| **Density:** | 9.32 g/cm³ (near room temperature) |
| **Appearance:** | Silvery-grey metal |

pure element in 1911. The process James carried out to procure this sample required 15,000 separate steps – proving that elemental thulium was almost as hard to find as the mythical land it was named after.

Like europium and terbium, thulium is sometimes used as a security measure in euro banknotes, in compounds that fluoresce with a blue colour under ultraviolet light. It can also be used in combination with holmium and chromium in yttrium–aluminium–garnet (YAG) lasers for military and medical applications. The isotope thulium-170 can act as a portable X-ray source. This isotope has a half-life of 128.6 days, and machines powered by thulium have a working life of approximately one year.

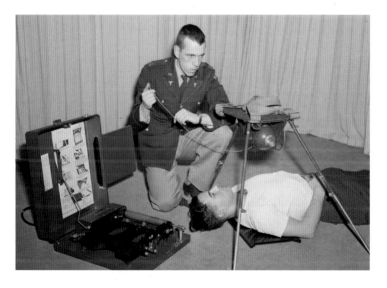

This portable X-ray device, photographed in 1955, was powered by radioactive thulium, which meant that it could be used in the field in contexts where no electricity, water supply or darkroom were available.

# Ytterbium

DISCOVERY DATE:1878     DISCOVERED BY: Jean Charles Galissard de Marignac

Ytterbium, a silvery-white metal in the lanthanide series, was the last of four elements to be named after the Swedish town of Ytterby, although this was not a straightforward decision. In fact, the debate about this element's name lasted for decades.

It started with the discovery in 1878 by Jean Charles Galissard de Marignac of a mineral earth that he named ytterbia. Then in 1905, Carl Auer von Welsbach described two new elements derived from ytterbia, which he called aldebaran and cassiopium. Two years later, Georges Urbain announced two new compounds in ytterbia and gave them the names neoytterbia and lutecia. Finally, in 1909, the credit for discovering these two elements was officially awarded to Georges Urbain, and they were named ytterbium (element 70) and lutetium (element 71).

Although Ytterbium was discovered in 1878, it was not until 1953 that a relatively pure sample of the metal was obtained.

**70**

# Yb

Ytterbium

173.045

| | |
|---|---|
| Atomic number: | 70 |
| Group: | Lanthanides |
| Period: | Period 6 |
| Block: | f-block |
| Atomic mass: | 173.045 u |

| | |
|---|---|
| Melting point: | 824 °C, 1515 °F |
| Boiling point: | 1196 °C, 2185 °F |
| Density: | 6.90 g/cm³ (near room temperature) |
| Appearance: | Silvery-white metal |

This ytterbium lattice atomic clock is, amazingly, up to ten times more accurate than its caesium-powered equivalents.

Ytterbium has a variety of industrial applications, including as a dopant to improve the qualities of stainless steel, as a source of gamma rays in portable X-ray equipment (like its neighbour, thulium) and as an industrial catalyst. It is also used in atomic clocks, powered by an optical lattice containing 10,000 ytterbium atoms cooled down to 10 millionths of a degree above absolute zero. These clocks are even more accurate than caesium-based atomic clocks, with an accuracy of only one second in the entire existence of the Universe.

One unusual property of ytterbium is that its electrical conductivity varies with pressure. This means it can be used in equipment to measure huge amounts of pressure, such as in earthquakes and nuclear explosions.

# Lutetium

DISCOVERY DATE: 1907 DISCOVERED BY: Georges Urbain (France), Carl Auer von Welsbach (Austria) & Charles James (USA)

Lutetium was the last of the lanthanides to be discovered, by Georges Urbain in France, Carl Auer von Welsbach in Austria and Charles James in the USA, in 1907. All three scientists independently identified this new element from its parent mineral, ytterbia, but Urbain was officially given credit for the discovery as he had published his findings first. Urbain's chosen name for this element was lutecium, after the Latin name for his home city of Paris, Lutetia, and this was formally adopted. In 1949, the official spelling was changed to lutetium, and this spelling remains in use today.

Lutetium is the heaviest, densest and hardest of all the lanthanide elements, and it has the highest melting point of the group. Some scientists place lutetium not in the lanthanides but in the transition metal group.

Lutetium, element 71, is a silvery white metal that is slightly more abundant in the Earth's crust than silver.

| | |
|---|---|
| **71** | |
| **Lu** | |
| Lutetium | |
| 174.967 | |

| | |
|---|---|
| Atomic number: | 71 |
| Group: | Lanthanides |
| Period: | Period 6 |
| Block: | f-block |
| Atomic mass: | 174.967 u |
| Melting point: | 1652 °C, 3006 °F |
| Boiling point: | 3402 °C, 6156 °F |
| Density: | 9.841 g/cm³ (near room temperature) |
| Appearance: | Silvery-white metal |

**Georges Urbain (1872–1938) was a French chemist who discovered lutetium and spent much of his career working on the lanthanide or rare earth elements.**

As lutetium is difficult to obtain in significant quantities, it only has a limited number of commercial applications. One of these is lutetium–hafnium dating, which is used to work out the age of meteorites and other rocks by measuring the decay of lutetium-176 into hafnium-176. This process has a half-life of 37.1 billion years, a fact that can be used to provide a timescale to assess the age of samples.

Lutetium oxide is used as a catalyst in the petrochemicals industry, to break down the large hydrocarbon molecules in crude oil into smaller products that are more useful. In medicine, the radioactive isotope lutetium-177 is used in cancer treatments for tumours in the nervous system or the endocrine system. It has also been used in trials as a treatment for advanced prostate cancer.

# Hafnium

DISCOVERY DATE: 1923    DISCOVERED BY: George Charles de Hevesy and Dirk Coster

Hafnium is a shiny, silvery-grey transition metal that is chemically very similar to the element immediately above it in the periodic table, zirconium. Dmitri Mendeleev had predicted the existence of the then-unknown element 72 in 1869, but hafnium was not discovered until 1923, by Dirk Coster and George Charles de Hevesy, in Copenhagen, Denmark. Hafnium's name comes from the Latin name for Copenhagen, Hafnia.

Hafnium resists corrosion and has a high melting point, which makes it useful in welding torches and plasma cutting tips. It is also used in the control rods in nuclear reactors, where it absorbs neutrons 600 times more effectively than zirconium. Hafnium's temperature resistance has led it to be used in some computer processors, and hafnium oxide is employed in the blue lasers of DVD readers, as well as in electrical insulators for microchips.

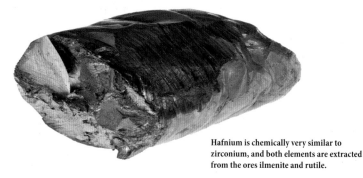

Hafnium is chemically very similar to zirconium, and both elements are extracted from the ores ilmenite and rutile.

| 72 | | |
|---|---|---|
| **Hf** | | |
| Hafnium | | |
| 178.486 | | |

| | |
|---|---|
| **Atomic number:** | 72 |
| **Group:** | Group 4 |
| **Period:** | Period 6 |
| **Block:** | d-block |
| **Atomic mass:** | 178.486 u |

| | |
|---|---|
| **Melting point:** | 2233 °C, 4051 °F |
| **Boiling point:** | 4603 °C, 8317 °F |
| **Density:** | 13.31 g/cm³ (near room temperature) |
| **Appearance:** | Steel-grey metal |

George de Hevesy
(1885–1966)
was a Hungarian
radiochemist and
co-discoverer
of the element
hafnium with
Dutch physicist
Dirk Coster.

Hafnium has no known biological function and is not generally considered to be toxic, but the American National Institute for Occupational Safety and Health has set a maximum TWA (time-weighted average) exposure limit of 0.5mg/m3 over an eight-hour workday. This level is judged to be immediately dangerous to life and health. Another possible danger of hafnium is that, in powder form, it can spontaneously catch fire when exposed to air. This means that special care is needed by anyone machining hafnium.

# Tantalum

DISCOVERY DATE: 1802     DISCOVERED BY: Anders Gustaf Ekeberg

Tantalum is a shiny blue-grey transition metal that is very hard, ductile and resistant to corrosion. It is named after Tantalus, a figure from Greek mythology who was punished for trying to trick the gods into eating his son by being made to stand forever in a pool of water by a tree with low-hanging fruit on its branches, but with both the water and the fruit always remaining just out of his reach. Anders Ekeberg, the Swedish chemist who discovered tantalum, gave it this name because when he immersed it in acid, it did not absorb any of the acid or become saturated by it.

Tantalum has a range of industrial applications. It is used in alloys to give strength and heat resistance, which makes it particularly useful for jet engine components and turbine blades. Another application is in electrical components for mobile phones and computers, especially in capacitors and resistors.

Tantalum, a highly corrosion-resistant metal, is chemically inert and has a very high melting point.

| 73 | | |
|---|---|---|
| **Ta** | | |
| Tantalum | | |
| 180.948 | | |

| | |
|---|---|
| Atomic number: | 73 |
| Group: | Group 5 |
| Period: | Period 6 |
| Block: | d-block |
| Atomic mass: | 180.948 u |

| | |
|---|---|
| Melting point: | 3017 °C, 5463 °F |
| Boiling point: | 5458 °C, 9856 °F |
| Density: | 16.69 g/cm³ (near room temperature) |
| Appearance: | Blue-grey metal |

154

Tantalum is not needed in the human body, but it does have an important medical use. Because it does not react with bodily fluids or provoke any immune response, it is ideally suited for use in dental implants and surgical instruments.

One of the most important sources of tantalum is an ore called coltan, which is also a source of niobium. Most coltan is found in central Africa, and the smuggling and export of this ore has been a key factor in prolonging and funding warfare in the Democratic Republic of Congo, in a conflict which has seen 5.4 million deaths since 1998.

Tantalum is widely used in the electronics industry, such as in the capacitors on this printed circuit board.

# Tungsten

DISCOVERY DATE: 1783    DISCOVERED BY: Juan and Fausto Elhuyar

Tungsten is a very hard, silver-grey metal that was identified in 1781 by Carl Wilhelm Scheele in Sweden and isolated as a metal in 1783 by the Spanish brothers Juan José and Fausto Elhuyar. It has the highest melting point of any metallic element (3422°C), and of all the elements, is second only to carbon in this respect.

Tungsten's name comes from a Swedish word meaning 'heavy stone', describing the mineral scheelite, an ore that contains this element. Its symbol, W, comes from an alternative name, wolfram, based on another tungsten-containing ore, wolframite. Intriguingly, this latter name comes from the German *wolf rahm*, meaning 'wolf soot' or 'wolf cream', referring to the fact that the process of extracting tungsten from the ore devoured vast amounts of tin with a voracious, wolf-like appetite.

This chunk of pale quartz contains dark crystals of the mineral wolframite, a tungsten ore.

| 74 **W** Tungsten 183.84 | | |
|---|---|---|
| **Atomic number:** | 74 | |
| **Group:** | Group 6 | |
| **Period:** | Period 6 | |
| **Block:** | d-block | |
| **Atomic mass:** | 183.84 u | |
| **Melting point:** | 3422 °C, 6192 °F | |
| **Boiling point:** | 5930 °C, 10706 °F | |
| **Density:** | 19.25 g/cm³ (near room temperature) | |
| **Appearance:** | Greyish-white metal | |

Classic light bulbs contain a tungsten filament surrounded by non-combustible inert gas, such as argon, to make sure that the heated wire will not catch fire.

The biggest application of tungsten, accounting for 55 per cent of its use, is in the production of cemented carbides, which are some of the hardest known substances in the world. These have a enormous range of applications across mining and construction, the aerospace and automotive industries, and electronics. Another 20 percent of global production goes into the manufacture of steels and other alloys. Although tungsten is well known for its use in light bulb filaments, only about 4 per cent of the global production of tungsten is directed towards this function.

Tungsten plays no role in human biology and it has the potential to cause cancers and other illnesses in many animals if inhaled as dust. However, it is present in enzymes in some species of bacteria and archaea, and it is actually essential for two archaea species.

157

# Rhenium

DISCOVERY DATE: 1925     DISCOVERED BY: Walter Noddack, Ida Tacke and Otto Berg

Rhenium, a hard silvery-grey transition metal, was the last stable element to be discovered, in 1925. It had, in fact, been discovered in 1908 by a Japanese chemist called Masataka Ogawa, but he wrongly placed it at position 43 in the periodic table (which would later be filled by technetium, in 1937). Therefore, the official discovery was awarded to the German chemists Walter Noddack, Ida Tacke and Otto Berg, 17 years later. They named the new element after the river Rhine, whose Latin name is Rhenus.

Rhenium is one of the rarest elements in the Earth's crust, and less than 50 tonnes (55 tons) of it are refined each year, with another 10 tonnes (11 tons) being reclaimed through recycling programmes. The main sources of rhenium are ores containing molybdenum and manganese, and production is primarily based in Chile, the USA,

Rhenium is usually produced in the form of a powder, but can be pressed and sintered into a solid form.

| 75 | |
|---|---|
| **Re** | |
| Rhenium | |
| 186.207 | |

| | |
|---|---|
| Atomic number: | 75 |
| Group: | Group 7 |
| Period: | Period 6 |
| Block: | d-block |
| Atomic mass: | 186.207 u |

| | |
|---|---|
| Melting point: | 3186 °C, 5767 °F |
| Boiling point: | 5630 °C, 10,170 °F |
| Density: | 21.02 g/cm³ (near room temperature) |
| Appearance: | Silvery-grey metal |

Re

Poland, Korea and China. A large percentage of rhenium production goes into the manufacture of high-temperature alloys for jet engine parts. Because these alloys can function at temperatures of up to 1600°C, they allow engines to operate more efficiently and with lower emissions of nitrous oxide ($N_2O$). Rhenium–platinum alloys are used as catalysts to convert crude oil products with low octane ratings into high-octane liquid products.

There is no known biological role played by rhenium, but two of its isotopes, rhenium-186 and rhenium-188, are used to treat liver cancer. The latter of these two isotopes is also used in the treatment of skin cancer and pancreatic cancer.

**Rhenium, with its very high melting point, is a valuable constituent of superalloys used in jet engines, such as this Airbus A350 Rolls-Royce Trent XWB engine.**

# Osmium

DISCOVERY DATE: 1803    DISCOVERED BY: Smithson Tennant and William Hyde Wollaston

Osmium is a hard, brittle, shiny metal and a member of the platinum group (with ruthenium, rhodium, palladium, iridium and, of course, platinum). It is one of the rarest elements in the Earth's crust, making up only 50 parts per trillion, and it is also the densest naturally occurring element, closely followed by iridium. Osmium was discovered in 1803 by Smithson Tennant and William Hyde Wollaston, and it was named by Tennant after the Greek word osme, meaning 'smell', because the osmium tetroxide he had produced gave off a smell reminiscent of chlorine and garlic.

Osmium's hardness and resistance to corrosion have led to it being used in a wide range of commercial applications, including fountain pen nibs, compass pointers and gramophone needles. Modern record players now tend to have styluses with diamond or sapphire tips, which are more expensive but last longer than their osmium-tipped

Osmium is the rarest precious metal, accounting for only 50 parts per trillion in the Earth's crust.

| 76 | |
|----|----|
| **Os** | |
| Osmium | |
| 190.23 | |

| | |
|---|---|
| **Atomic number:** | 76 |
| **Group:** | Group 8 |
| **Period:** | Period 6 |
| **Block:** | d-block |
| **Atomic mass:** | 190.23 u |
| **Melting point:** | 3033 °C, 5491 °F |
| **Boiling point:** | 5012 °C, 9054 °F |
| **Density:** | 22.59 g/cm³ (near room temperature) |
| **Appearance:** | Shiny silver metal |

predecessors. Osmium was also once used in light bulb filaments, although it was replaced in this function by tungsten. The German light bulb manufacturer Osram based its name on osmium and wolfram (another name for tungsten), highlighting just how important these two elements have been in bringing light to the world.

Osmium tetroxide forms strong bonds with fats and oils, leaving a black residue of osmium dioxide, and it was once used by detectives to reveal fingerprints, although this use came to an end when the toxicity of this compound was confirmed. It is, however, still used in medical research as a stain for biological samples, making them easier to view under an electron microscope.

Osmium was used for the stylus tips of record players in the 1950s and 1960s, before these were gradually replaced by sapphire and diamond tips.

# Iridium

DISCOVERY DATE: 1803     DISCOVERED BY: Smithson Tennant

Iridium is very rare, but an unusually high amount of it can be found in a fine layer of quartz dust in rocks all over the world. This dates from 65 million years ago, when a 10-km (6.2-mile)-wide asteroid crash-landed in Mexico, causing a vast explosion that is believed to have led to the extinction of dinosaurs on Earth. The layer of iridium-containing dust is the remaining debris from that impact.

Iridium is the second-densest metal that exists naturally on Earth, after osmium. It is resistant to corrosion by air, water and acids, and only molten sodium cyanide or potassium cyanide can dissolve it. The British chemist who discovered iridium in 1803, Smithson Tennant, named it after Iris, the Greek goddess of the rainbow, because its salts are so brightly coloured.

Iridium is a hard, brittle, silver-coloured metal named after the Greek goddess of the rainbow due to its colourful salts.

| 77 | |
|---|---|
| **Ir** | |
| Iridium | |
| 192.217 | |

| | |
|---|---|
| **Atomic number:** | 77 |
| **Group:** | Group 9 |
| **Period:** | Period 6 |
| **Block:** | d-block |
| **Atomic mass:** | 192.217 u |
| **Melting point:** | 2446 °C, 4435 °F |
| **Boiling point:** | 4130 °C, 7466 °F |
| **Density:** | 22.56 g/cm³ (near room temperature) |
| **Appearance:** | Silvery-white metal |

Supplies of iridium are very limited due to its scarcity, with fewer than 10 tonnes (11 tons) being produced every year and a 2024 market value of US$160 per gram. However, despite being in short supply, iridium has a number of key industrial applications. Its hardness and high melting point make it valuable for use in the tips of spark plugs and in crucibles. Iridium-osmium alloys are used for compass bearings and fountain pen nibs, and an alloy of iridium and titanium is used to produce deep-water pipes.

Iridium, often alloyed with osmium, adds a touch of style to a fountain pen nib.

# Platinum

DISCOVERY DATE: 1735     DISCOVERED BY: Antonio de Ulloa

Platinum is a shiny silvery-white metal that is ductile, malleable and highly unreactive. It also resists corrosion, even at high temperatures. Its name comes from the Spanish word platina, a diminutive form of *plata*, and means 'little silver'.

Platinum has been used since ancient times, with the earliest known artefact being a box from an Egyptian burial decorated with hieroglyphics in a gold-platinum alloy, dating back to 1200 BCE. It was mined and prized by Indigenous peoples in Latin America, and in 1735, a Spanish naval officer and scientist became the first European to systematically study and describe the metal, with the result that history has recorded his name as its discoverer.

Today, the country with the largest production of platinum is South Africa, followed by Russia, Zimbabwe, Canada and the USA. The main industrial application

Platinum is one of the least reactive metals, and one of the rarest elements in the Earth's crust.

| 78 | |
|----|--|
| **Pt** | |
| Platinum | |
| 195.084 | |

| | |
|---|---|
| **Atomic number:** | 78 |
| **Group:** | Group 10 |
| **Period:** | Period 6 |
| **Block:** | d-block |
| **Atomic mass:** | 195.084 u |
| **Melting point:** | 1768.3 °C, 3214.9 °F |
| **Boiling point:** | 3825 °C, 6917 °F |
| **Density:** | 21.45 g/cm³ (near room temperature) |
| **Appearance:** | Silvery-white metal |

**One-third of all the platinum sold in 2014 was used in jewellery, which demonstrates just how popular this precious silver-coloured metal is, as an adornment and an investment.**

of this precious metal is in catalytic converters for cars, which convert unburned hydrocarbons in exhaust gas into harmless water vapour and carbon dioxide. Platinum is also used for jewellery and in watch-making, where it is valued for the fact that it will not tarnish or wear out, being harder than gold.

Because platinum is so stable and long-lasting, it has been used to define standards of measurement. Between 1889 and 1960, the metre was defined by a bar of 90 per cent platinum and 10 per cent iridium at the melting point of ice. (For its current definition, see the Krypton entry in this book.) The same alloy was also used to define the weight of a kilogram from 1889 until 2019, when this was replaced by a new definition based on physical constants.

# Gold

DISCOVERY DATE: Approximately 3000 BCE

Gold must be one of the most famous and widely desired elements in the periodic table, due to its shiny yellow appearance. It is one of the least reactive elements and does not corrode, meaning that it holds its value, and it can be found in its pure elemental form or mixed with other precious metals in rocks and river beds. Gold's chemical symbol, Au, comes from the Latin word *aurum*, which describes the golden glow of sunrise. As well as being beautiful, gold is also the most ductile and most malleable of all the elements. This means it can be pulled into fine wire, and hammered into foils so thin that they are actually transparent.

Estimates of how much gold has been excavated from the Earth range from 178,100 tonnes (196,211 tons) to 212,582 tonnes (234,331 tons) and an estimated 59,000 tonnes (65,036 tons) remain. Another 20 million tonnes (22 million tons) of gold are present in

Gold often occurs in a pure elemental state as it is too unreactive to easily form compounds with other elements. This nugget of gold and quartz was found in Nevada, USA.

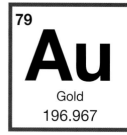

**79**

**Au**

Gold

196.967

| | |
|---|---|
| Atomic number: | 79 |
| Group: | Group 11 |
| Period: | Period 6 |
| Block: | d-block |
| Atomic mass: | 196.967 u |

| | |
|---|---|
| Melting point: | 1064.18 °C, 1947.52 °F |
| Boiling point: | 2970 °C, 5378 °F |
| Density: | 19.3 g/cm³ (near room temperature) |
| Appearance: | Shiny yellow metal |

The funerary mask of the Egyptian pharaoh Tutankhamun, who reigned from 1334 to 1325 BCE, was lavishly decorated with gold and precious jewels.

the world's oceans in a diluted form, and people have long researched ways of extracting it. To date, however, no practical method has been found.

Gold is used widely in jewellery, and it is held as gold bullion bars by banks as a solid form of wealth with a relatively stable value. Although many monetary currencies were once defined in terms of the value of a specific quantity of gold, most countries moved away from this system after World War II, and Switzerland was the last country to leave the gold standard in 1999.

Because it is so inert, gold leaf can be eaten safely without being absorbed into the bloodstream. Gold leaf can be found as a decoration on luxury chocolates and foods, and even in vodka and cinnamon schnapps.

# Mercury

DISCOVERY DATE: Approximately 1500 BCE

Mercury is one of only two elements that are liquid at standard room temperature and pressure, the other being bromine, element number 35. Its name comes from Roman mythology, where Mercury was the messenger of the gods, travelling swiftly with wings on his hat and his sandals, and reflects the flowing metal's ability to move just as quickly. Another English name for mercury is 'quicksilver', and its chemical symbol, Hg, comes from the Latin word *hydrargyrus*, meaning 'water silver'.

In the past, mercury has had a wide variety of industrial applications, because of its ability to form amalgams with other metals, making it easier to work with them at lower temperatures. These applications have included use in batteries, dental fillings, mirrors, cosmetics and fluorescent lighting, but many of these have been phased out due to mercury's dangerous toxicity. Mercury is still used in the chemical industry

Mercury is the only metal to be liquid at standard room temperature and pressure.

| 80 | |
|---|---|
| **Hg** | |
| Mercury | |
| 200.592 | |

| | |
|---|---|
| Atomic number: | 80 |
| Group: | Group 12 |
| Period: | Period 6 |
| Block: | d-block |
| Atomic mass: | 200.592 u |

| | |
|---|---|
| Melting point: | −38.8290 °C, −37.8922 °F |
| Boiling point: | 356.73 °C, 674.11 °F |
| Density: | 13.546 g/cm³ (near room temperature) |
| Appearance: | Silver-coloured liquid |

as a catalyst, and in glass thermometers, particularly those used for measuring very high temperatures.

The toxic effects of mercury have affected people for centuries, whether through workplace exposure or the result of industrial pollution. The Mad Hatter in the 1865 book *Alice's Adventures in Wonderland* was not merely a creation of Lewis Carroll's imagination but had his origin in the fact that many textile workers, including hatters, used to suffer disorders of the nervous system after breathing in mercury vapours.

A small amount of mercury is present in every living thing, and also in the food we eat. Most of our exposure to mercury comes through ingesting the methylmercury in the flesh of various types of fish. Although the amount absorbed is not generally harmful, pregnant people are advised to limit their fish consumption to avoid this exposure.

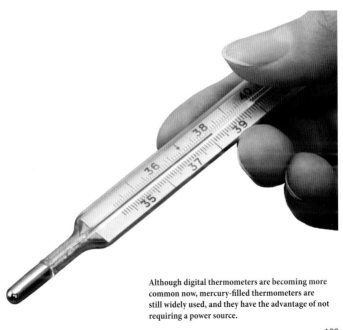

Although digital thermometers are becoming more common now, mercury-filled thermometers are still widely used, and they have the advantage of not requiring a power source.

# Thallium

DISCOVERY DATE: 1861     DISCOVERED BY: William Crookes

Thallium is a silvery-white post-transition metal that tarnishes when exposed to air. Its name comes from the Greek word *thallós*, meaning 'twig' or 'green shoot', because of the bright green spectral emission lines it produces. It was discovered independently by William Crookes in England in 1861 and by Claude-Auguste Lamy in France in 1862, with Lamy being the first of the two scientists to produce a pure sample of the element.

Thallium is extremely toxic, because its positively charged ions are almost identical to potassium ions, and when incorporated into the human body in place of potassium, it damages the nervous system, causing hair loss, nausea, constipation, increased heart rate, and eventually, complete heart failure and death. Thallium sulfate was once easily obtainable as a rat killer, and, being almost impossible to detect, became known as the 'poisoner's poison'. A dose of just 15 millions per kilogram of bodyweight is sufficient to kill someone.

This rod of thallium shows how the metal corrodes when exposed to air.

| 81 | |
|---|---|
| **Tl** | |
| Thallium | |
| 204.38 | |

| | |
|---|---|
| Atomic number: | 81 |
| Group: | Group 13 |
| Period: | Period 6 |
| Block: | p-block |
| Atomic mass: | 204.38 u |

| | |
|---|---|
| Melting point: | 304 °C, 579 °F |
| Boiling point: | 1473 °C, 2683 °F |
| Density: | 11.85 g/cm³ (near room temperature) |
| Appearance: | Silvery-white metal |

William Crookes (1832–1919), the discoverer of thallium, was a pioneer of vacuum tubes for the study of physical phenomena, and invented the Crookes tube. He discovered that the rays (which were made of electrons) made the sides of the glass tube fluorescent.

Thallium has played its role in fiction, as the murder weapon in Agatha Christie's mystery novel *The Pale Horse*, published in 1961. Amazingly, this novel helped readers identify thallium poisoning before it was too late in at least two cases, and in 1971, a British serial killer called Graham Frederick Young was caught because a doctor working with Scotland Yard recognized the symptoms of a mysterious 'illness' as thallium poisoning after reading the book.

Despite its toxicity, thallium has some important industrial applications. It is combined with mercury for use in thermometers designed for measuring low temperatures, because while mercury freezes at -39°C (-38.2°F), a thermometer containing both metals can measure temperatures as low as -60°C (-76°F). It is also used in electronics, glass-making and nuclear medicine.

# Lead

Lead is a soft, malleable metal with a shiny grey surface that rapidly oxidizes in air to a dull matt finish. It is the last truly stable element in the periodic table: from element 83 onwards, all elements are unstable, which means that they decay over time – or, in other words, that they are radioactive.

Lead has been known since antiquity and its Latin name, *plumbum*, gives us lead's chemical symbol, Pb. This Latin root created several words in English, including 'plumbing', 'plumber' and 'plumb line', reflecting just how important lead has been throughout our history.

Due to it being easy to shape and resistant to corrosion, lead has been used in the manufacture of pipes, paints and pottery glazes for many centuries, and it was even used as a cosmetic, known as Venetian ceruse, in the eighteenth century. However, in all these

The yellow crystal known as anglesite is a lead sulfate mineral, with the chemical formula $PbSO_4$.

| | |
|---|---|
| **82** | |
| **Pb** | |
| Lead | |
| 207.2 | |

| | |
|---|---|
| Atomic number: | 82 |
| Group: | Group 14 |
| Period: | Period 6 |
| Block: | p-block |
| Atomic mass: | 207.2 u |

| | |
|---|---|
| Melting point: | 327.46 °C, 621.43 °F |
| Boiling point: | 1749 °C, 3180 °F |
| Density: | 11.34 g/cm³ (near room temperature) |
| Appearance: | Dull grey metal |

applications, lead was dangerously toxic to humans, with the potential to damage almost every system and organ in the body. As a result, it has been largely phased out, although it was only towards the end of the twentieth century that lead started to be removed from petrol, where it had functioned as an antiknock agent. Algeria was the last country to ban the use of lead in petrol, in 2021.

Lead pollution is still a significant problem today and lead poisoning is estimated to make up 1 per cent of the world's total disease burden, with a particular impact on brain development in children. Despite this, there are still a number of applications for lead, in contexts where it is less likely to cause harm. These include lead-acid car batteries, diving belts, organ pipes and protective shields and aprons for use in X-ray rooms.

Lead has been used since ancient times. Its malleability and resistance to corrosion make it ideal for use in stained-glass windows.

# Bismuth

Bismuth is a brittle, silvery-white metal that has been known since ancient times. Its name dates back to around 1665 and may be derived from the German word *Wismuth*, which can itself be traced to the phrase *weisse masse* or 'white mass'. For nearly the entire period that bismuth has been known to humanity, it was thought to be a stable element, but in 2003 scientists discovered that it was, in fact, very weakly radioactive. Its primordial isotope (a form that dates back to before the formation of the Earth), bismuth-209, has an incredibly long half-life of 19 billion billion years – more than a billion times longer than the age of the Universe.

Bismuth is almost as dense as lead, and has a low melting point. As a result, bismuth alloys are sometimes used as a replacement for lead, which has been phased out due

**Pure bismuth is rarely found in nature, but when synthesised can take the distinctive form of these hopper crystals.**

| 83 | | |
|---|---|---|
| **Bi** | | |
| Bismuth | | |
| 208.98 | | |

| | |
|---|---|
| **Atomic number:** | 83 |
| **Group:** | Group 15 |
| **Period:** | Period 6 |
| **Block:** | p-block |
| **Atomic mass:** | 208.98 u |

| | |
|---|---|
| **Melting point:** | 271.5 °C, 520.7 °F |
| **Boiling point:** | 1564 °C, 2847 °F |
| **Density:** | 9.78 g/cm³ (near room temperature) |
| **Appearance:** | Crystalline silvery metal |

**Bismuth oxychloride (BiOCl) gives a pearly, iridescent finish to nail polish and other cosmetics.**

to its toxic effects. Examples of these applications include its use in fishing weights, shot for the hunting of birds, shields for use during X-rays and solders. Compounds containing bismuth have a surprisingly wide range of uses. Bismuth oxychloride is used in the cosmetics industry to give a luminescent, pearly shine to make-up products. Bismuth(III) oxide helps to make celebrations go with a bang by creating the crackling sound in 'Dragon's Egg' fireworks. And finally, for anyone who has celebrated a little too enthusiastically, bismuth subsalicylate is the active ingredient in 'pink bismuth' products such as Pepto-Bismol, a remedy for diarrhoea, nausea and indigestion.

# Polonium

DISCOVERY DATE: 1898    DISCOVERED BY: Marie Skłodowska-Curie

Polonium is a grey metal that is highly radioactive. All of its isotopes have relatively short half-lives, making it one of the 10 rarest elements on Earth.

Polonium was discovered in France by Marie Skłodowska-Curie and her husband Pierre Curie. They had extracted uranium and thorium from a uranium ore known as pitchblende, and noticed that the remaining material was even more radioactive than the elements that had been removed. After identifying polonium in July 1898, they went on to discover another element in pitchblende five months later: radium. Polonium was named after Skłodowska-Curie's homeland, Poland, which at this time had been divided between Russia, Germany and the Austro-Hungarian Empire. The name was chosen to raise awareness of Poland's status, and polonium was therefore the first element to be explicitly named with a political purpose.

Uraninite, or pitchblende, was the source for a number of elemental discoveries, including polonium.

| 84 Po | |
|---|---|
| Polonium | |
| [208.98] | |

| | |
|---|---|
| **Atomic number:** | 84 |
| **Group:** | Group 16 |
| **Period:** | Period 6 |
| **Block:** | p-block |
| **Atomic mass:** | [208.98] |
| **Melting point:** | 254 °C, 489 °F |
| **Boiling point:** | 962 °C, 1764 °F |
| **Density:** | 9.196 g/cm³ (alpha, near room temperature) |
| | 9.398 g/cm³ (beta, near room temperature) |
| **Appearance:** | Silvery-grey semi-metal |

Due to its extreme radioactivity, polonium's applications are limited. It can be used in antistatic brushes used to remove dust particles from photographic film, and to eliminate static charges in textile mills, but safer beta-particle sources have often replaced polonium in these roles. It can also provide an atomic heat source for moon rovers and satellites.

Sadly, polonium is well known for its deliberate use as a radioactive poison. One very high-profile case was the assassination of Alexander Litvinenko, a Russian former FSB officer who had defected to the UK, in 2006. When Litvinenko fell ill in November 2006 and was admitted to hospital in London, he was found to have been poisoned with polonium-210. He died 23 days later. A public enquiry conducted in 2015–16 concluded that his killing was probably carried out with the approval of Vladimir Putin.

Marie Skłodowska-Curie (1867–1934) was the first woman to receive a Nobel Prize, and the first person to receive two Nobel Prizes. The achievements she and her husband Pierre made were later commemorated in the name of the element curium.

# Astatine

DISCOVERY DATE: 1940     DISCOVERED BY: Dale R. Corson, Kenneth Ross MacKenzie and Emilio Segrè

Astatine is one of the many elements whose existence was predicted by Dmitry Mendeleev when he drew up his periodic table, but it proved extremely hard to find. Very small quantities of astatine are produced on Earth through the decay of naturally occurring uranium and thorium, but the total amount on the whole planet at any moment is estimated to be less than 30g (1oz).

Astatine was produced artificially for the first time in 1940 by Dale R. Corson, Kenneth Ross MacKenzie and Emilio Segrè at the University of California, Berkeley. They used a cyclotron to bombard bismuth-209 with alpha particles (helium nuclei), to produce astatine-211 and two free neutrons. The team of discoverers published their suggested name for the new element in a letter to the scientific journal *Nature* in 1947: astatine, based on the Greek word *astatos*, meaning 'unstable'.

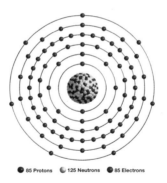

● 85 Protons  ◐ 125 Neutrons  ● 85 Electrons

This diagram shows the atomic structure of astatine, with red protons, yellow neutrons and shells of blue electrons.

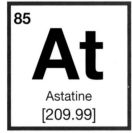

**85**

**At**

Astatine

[209.99]

| | |
|---|---|
| **Atomic number:** | 85 |
| **Group:** | Group 17 |
| **Period:** | Period 6 |
| **Block:** | p-block |
| **Atomic mass:** | [209.99] |
| **Melting point:** | 300 °C, 572 °F |
| **Boiling point:** | 350 °C, 662 °F |
| **Density:** | 8.91–8.95 g/cm³ (near room temperature, estimated) |
| **Appearance:** | Unknown |

**Emilio Segrè (1905–1989) was an Italian– American physicist who co-discovered technetium and astatine.**

All the isotopes of astatine are unstable, with astatine-210 having the longest half-life, at 8.1 hours. Because only tiny quantities have ever been produced, and these have not existed for very long, researchers have been unable to determine astatine's atomic mass or its density, and we may never know these exactly. If enough astatine existed to be visible to the naked eye, it is thought that it might be a black solid, based on its position in the halogen group, but it is so unstable that any such sample would immediately be vaporized by the heat of its own radioactivity.

Astatine has no industrial applications, because it is so scarce that there is not enough of it to use for anything. It also plays no biological function, but it is being researched for its potential use in nuclear medicine.

# Radon

DISCOVERY DATE: 1900     DISCOVERED BY: Friedrich Ernst Dorn

Radon is a colourless, odourless and tasteless gas, and the heaviest member of the noble gas group of the periodic table. Its name comes from radium, which emits radon as it decays, and it is formed naturally through the decay of uranium (which also produces radium). It was discovered by Friedrich Ernst Dorn in Halle, Germany in 1900, and in 1912 it was given the name niton, with the chemical symbol Nt, based on the Latin *nitens*, meaning 'shining', to describe the way that it glows at very low temperatures. Its current name, radon, has been in official use since 1923.

Radon is emitted naturally from deposits of granite and limestone, and is a dangerous source of radioactivity for people working in mines. In 1530, the Swiss philosopher Paracelsus described a wasting disease that affected miners, and in 1879, this illness was confirmed to be lung cancer by researchers in Germany. Radon

Radon, the heaviest of the noble gases, is dangerously radioactive and prolonged exposure to it can cause lung cancer.

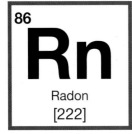

| 86 | |
|---|---|
| **Rn** | |
| Radon | |
| [222] | |

| | |
|---|---|
| Atomic number: | 86 |
| Group: | Group 18 |
| Period: | Period 6 |
| Block: | p-block |
| Atomic mass: | [222] |

| | |
|---|---|
| Melting point: | −71 °C, −96 °F |
| Boiling point: | −61.7 °C, −79.1 °F |
| Density: | 0.00973 g/cm³ (near room temperature) |
| Appearance: | Colourless gas |

**Rn**

The village of Furnes in the Azores is located in the caldera of an active volcano, and a 1999 study showed that hazardous levels of radon could affect about one-third of the village's inhabitants.

is also a danger in the basements of homes built on radon-rich land. Although the amount of radiation emitted daily might be small, in confined underground spaces it can accumulate, creating a serious health risk. To overcome this issue, ventilation systems are available to stop radon from entering living spaces.

Radon plays no biological role in the human body, although as a toxin, it is thought to be the main cause of lung cancer in non-smokers. In the past radon was widely – and wrongly – promoted as a health tonic, and in the early twentieth century some practitioners recommended the use of a 'radiotorium', a sealed room where patients could breathe radon for its supposed health benefits. There are still spas today where guests can drink and bathe in radon-rich water. Thankfully, a 1999 investigation by the US National Research Council showed that the risk associated with ingesting radon was almost zero, meaning that spa visitors should emerge from their treatments unscathed.

181

# Francium

DISCOVERY DATE: 1939     DISCOVERED BY: Marguerite Perey

Francium is the second least abundant element in the Earth's crust, after astatine, and it is thought that less than 30g (1oz) of francium exist on our planet at any given moment. It is the most unstable of all the naturally occurring elements and one of the most unstable of the elements that have been synthesized in laboratories. The longest half-life of any of its isotopes, francium-223, is 22 minutes.

Francium was the last of the naturally occurring elements to be discovered, by Marguerite Perey in Paris in 1939. Perey proposed the name 'catium' for it, with the chemical symbol Ct, to reflect her assessment that it was the most electro-positive of the elements, but this name was rejected by her supervisor, Irène Joliot-Curie. Instead, they opted for a name that celebrated Perey's homeland of France, even though a previously discovered element, gallium, had already been named in honour of the same nation.

This sample of the uranium ore uraninite comes from Příbram in the Czech Republic. Uraninite contains a ratio of one atom of francium to every $1 \times 10^{18}$ atoms of uranium.

| 87 | |
|---|---|
| **Fr** | |
| Francium | |
| [223.02] | |

| | |
|---|---|
| **Atomic number:** | 87 |
| **Group:** | Group 1 |
| **Period:** | Period 7 |
| **Block:** | s-block |
| **Atomic mass:** | [223.02] |
| **Melting point:** | 27 °C, 81 °F |
| **Boiling point:** | 677 °C, 1251 °F |
| **Density:** | 2.48 g/cm³ (near room temperature, estimated) |
| **Appearance:** | Unconfirmed; possibly a soft grey metal |

**Fr**

Marguerite Perey (1909–1975), the discoverer of francium, was a student of Marie Curie and the first woman to be elected to the French Académie des Sciences.

As the final element in Group 1 of the periodic table, francium might be expected to be the most reactive of all of them, but due to its heaviness, its electrons orbit its nucleus so quickly that they actually move closer to the centre of the atom, and therefore become less reactive. This means that francium is likely to be less reactive than the element directly above it in Group 1, caesium. However, due to its scarcity, it is unlikely that this will ever be directly put to the test. Like astatine, francium has never had any industrial applications outside of scientific research labs, as not enough of it has ever existed to use.

# Radium

DISCOVERY DATE: 1898 DISCOVERED BY: Marie Skłodowska-Curie and Pierre Curie

Radium is a highly radioactive, silvery-white metal that quickly develops a black oxidized coating when exposed to air. All of its isotopes are radioactive, and the most stable of these is radium-226, with a half-life of 1600 years.

Radium was discovered in December 1898 by Marie Skłodowska-Curie and Pierre Curie in a sample of pitchblende, an ore containing uranium. They had previously extracted uranium from the sample, and discovered that the remaining material was still radioactive. In July, they isolated a new element, polonium, from the sample, and then in December they identified a compound containing another new element. They named this new element radium from the Latin word *radius*, meaning 'ray', to describe the visible energy rays it emitted. At the time of this research, the dangers of radioactivity were not sufficiently understood, and Skłodowska-Curie herself absorbed

Radium, like many other radioactive elements, was extracted from uraninite, by Marie Skłodowska-Curie and Pierre Curie.

| 88 | |
|---|---|
| **Ra** | |
| Radium | |
| [226.03] | |

| | |
|---|---|
| Atomic number: | 88 |
| Group: | Group 2 |
| Period: | Period 7 |
| Block: | s-block |
| Atomic mass: | [226.03] |

| | |
|---|---|
| Melting point: | 700 °C, 1292 °F or 960 °C, 1760 °F |
| Boiling point: | 1737 °C, 3159 °F |
| Density: | 5.5 g/cm³ (near room temperature) |
| Appearance: | Silvery-white metal |

far more radiation than was safe. Her death in 1934 was caused by aplastic anaemia, which was believed to be the result of her long exposure to radiation. Even today, her laboratory notebooks are still radioactive, and they are stored in lead-lined boxes.

Radium had a wide range of commercial applications, including toothpaste and chocolate, before it was known to be dangerous. It was famously used to make the hands and numbers on watch dials glow in the dark, and many of the women who painted these dials suffered from anaemia and bone cancer after being instructed to lick the tips of the paint brushes to produce a fine point. It took years of protracted legal battles before these workers obtained any compensation. Today, the only commercial use of radiation is in nuclear medicine.

**This photo, taken in 1932, shows a laboratory assistant holding a test tube containing 1 gram of radium, using a lead screen to protect herself from its radioactivity.**

# Actinium

DISCOVERY DATE: 1899    DISCOVERED BY: André-Louis Debierne

Actinium is a silvery-white metal that glows with a blue light due to the radioactivity it emits. On exposure to air, it quickly forms an oxidized coating that prevents it from oxidizing further. All of its isotopes are radioactive, and the most stable of these has a half-life of roughly 22 years. It was discovered in 1899 by André-Louis Debierne, who found it while investigating the residues of a pitchblende sample after Marie Skłodowska-Curie and Pierre Curie had extracted radium from it. A German chemist, Friedrich Oskar Giesel, independently played a key role in this element's discovery, but Debierne's choice of name for the new element, actinium, is the one that has become officially accepted, based on the Greek word *aktinos*, meaning 'beam' or 'ray'.

Actinium is the first element in the actinide group, which fits neatly below the lanthanide group at the bottom of the periodic table. If these two groups took their

Actinium-225 medical radioisotope held in a v-vial. The blue glow comes from the ionization of surrounding air by alpha particles.

| 89 | |
|---|---|
| **Ac** | |
| Actinium | |
| [227] | |

| | |
|---|---|
| **Atomic number:** | 89 |
| **Group:** | Actinides |
| **Period:** | Period 7 |
| **Block:** | d-block |
| **Atomic mass:** | [227] |

| | |
|---|---|
| **Melting point:** | 1227 °C, 2240 °F (estimated) |
| **Boiling point:** | 3200±300 °C, 5800±500 °F (extrapolated) |
| **Density:** | 10 g/cm³ (near room temperature) |
| **Appearance:** | Soft silvery-white metal |

Friedrich Oskar Giesel (1852–1927) was the first scientist to successfully isolate the new element 89, which he named emanium. This same element was given the name actinium by André-Louis Debierne.

'correct' place in the table, it would be too wide to fit on the average display or poster, which is why these two groups are nearly always relegated to a separate zone. One key difference between these two outlying groups is that, while the lanthanides all have such similar properties that they can be hard to tell apart, the actinides all exhibit notably distinct characteristics.

There are not many commercial applications of actinium, due to its scarcity. It is used as a neutron source for laboratory purposes, and in neutron probes that measure the amount of water in soil. The isotope actinium-225 is being researched for possible future use in radiation therapy for certain cancers.

# Thorium

DISCOVERY DATE: 1829    DISCOVERED BY: Jöns Jacob Berzelius

Thorium, the second member of the actinide group, is a weakly radioactive silvery metal that tarnishes with an olive-coloured surface to form thorium dioxide. It is relatively soft and malleable, and it has the unusual quality of being pyrophoric, which means that it can ignite spontaneously in air.

Thorium is the most common of all the actinide elements, being around three times more abundant than uranium. The Earth's crust contains approximately the same amount of thorium as it does lead. It was identified as a new element by Swedish chemist Jöns Jacob Berzelius in 1829, after he had been sent a specimen of an unusual black mineral that had been discovered on Løvøya island, Norway by a Norwegian priest called Morten Thrane Esmark the previous year. Berzelius named the new element after the Norse god of thunder, Thor.

The mineral monazite is the main ore for the commercial production of thorium.

| 90 | |
|---|---|
| **Th** | |
| Thorium | |
| 232.038 | |

| | |
|---|---|
| Atomic number: | 90 |
| Group: | Actinides |
| Period: | Period 7 |
| Block: | f-block |
| Atomic mass: | 232.038 u |

| | |
|---|---|
| Melting point: | 1750 °C, 3182 °F |
| Boiling point: | 4788 °C, 8650 °F |
| Density: | 11.7 g/cm³ (near room temperature) |
| Appearance: | Silvery metal |

Several nuclear power plants with thorium reactors have been built, including this one in Hamm-Uentrop district of the city of Hamm in Germany.

Th

Thorium had a wide range of industrial applications before its radioactivity was understood, including in ceramics, carbon arc lamps and as an industrial catalyst. Thorium dioxide was used in the mantles of gas-powered streetlights because it glows when heated. It is actually still used for this purpose in some camping lamps today, because the glass casing provides enough protection for people using the lamps. Thorium oxide is also used for crucibles because of its very high melting point, which is the highest of any oxide. A possible future use of thorium is as an alternative fuel to uranium in nuclear reactors.

# Protactinium

DISCOVERY DATE: 1913     DISCOVERED BY: Kasimir Fajans (USA) and Oswald Göhring (Germany)

Protactinium is a dense and radioactive metal with a lustrous silvery-grey colour. It was discovered in 1913 by the Polish American chemist Kasimir Fajans and his German colleague Oswald Göhring in Karlsruhe, Germany. The two scientists suggested the name 'brevium' for the new element due to the short half-life of the specific isotope they were studying, protactium-234m.

In 1917–18, the Austrian–Swedish physicist Lise Meitner discovered a more stable isotope, protactinium-231, which has a half-life of 32,760 years. This is the most stable of the 29 isotopes that have been discovered, and accounts for almost all of the protactinium found on Earth. Meitner and her collaborator Otto Hahn suggested the name 'proto-actinium' for their discovery, to note the fact that this element is the 'parent' of actinium, element 89, through radioactive decay. The element's name was formalized

This sample of the mineral torbernite may contain one or two atoms or protactinium. The element itself occurs in such small amounts that it is difficult to photograph in its pure form.

| 91 |
|---|
| **Pa** |
| Protactinium |
| 231.036 |

| | |
|---|---|
| Atomic number: | 91 |
| Group: | Actinides |
| Period: | Period 7 |
| Block: | f-block |
| Atomic mass: | 231.036 u |

| | |
|---|---|
| Melting point: | 1568 °C, 2854 °F |
| Boiling point: | 4027 °C, 7280 °F |
| Density: | 15.37 g/cm³ (near room temperature) |
| Appearance: | Silvery metal |

Protactinium was not identified as a new element until 1913, by Kasimir Fajans (pictured) and Oswald Göhring.

as the slightly easier to pronounce 'protactinium' by the International Union of Pure and Applied Chemistry in 1949.

Protactinium does not have many applications, and nor does it possess any particular remarkable qualities. As a result, it was very nearly named the most boring element by the scientific journal *Nature* in 2019. However, it is used in scientific research as a tracer in the radiometric dating of sediments up to 175,000 years old. Scientists can calculate the age of a sample by measuring the ratio of protactinium-231 to thorium-230 contained in it. Although protactinium is toxic and radioactive, a trace amount of it is – somewhat surprisingly – present in many homes. This is because the americium-241 which is fundamental to the operation of domestic smoke detectors decays over time into neptunium-237 and then into protactinium-233.

# Uranium

DISCOVERY DATE: 1789     DISCOVERED BY: Martin Heinrich Klaproth

Uranium is a silver-grey metal in the actinide series, and the heaviest element to exist on Earth without being created in a laboratory. Two of the 28 known isotopes of uranium occur naturally, and of this, 99 per cent is the isotope uranium-238, which has a half-life of 4.47 billion years. Most of the remaining amount is made up of uranium-235, with a half-life of 704 million years. Uranium-235 is unique in being the only naturally occurring nuclide that is fissile, meaning that it can sustain a nuclear chain reaction.

The discovery of uranium is credited to the German chemist Martin Heinrich Klaproth. In 1789, Klaproth dissolved pitchblende in nitric acid and heated the resulting yellow compound with charcoal to produce a black powder. This turned out to be uranium oxide, although at the time he thought it was pure uranium. He named the new element after the planet Uranus, which had been discovered in 1781. In 1841,

The mineral autunite, which was discovered in 1852, has a uranium content of 48.27 per cent and glows with a green light under a UV lamp.

| 92 | |
|---|---|
| **U** | |
| Uranium | |
| 238.0289 | |

| | |
|---|---|
| Atomic number: | 92 |
| Group: | Actinides |
| Period: | Period 7 |
| Block: | f-block |
| Atomic mass: | 238.0289 u |

| | |
|---|---|
| Melting point: | 1132.2 °C, 2070 °F |
| Boiling point: | 4131 °C, 7468 °F |
| Density: | 19.1 g/cm³ (near room temperature) |
| Appearance: | Silvery metal |

The Rössing Uranium Mine in Namibia is one of the world's largest open pit uranium mines, and has been running since 1976.

Eugène Péligot succeeded in isolating a sample of pure uranium metal, and then in 1896, Henri Becquerel discovered that uranium is radioactive, after leaving a sample of the metal on a photographic plate and observing that the plate became cloudy.

One of the most important applications of uranium is as a fuel in nuclear power plants, where it is capable of producing 1.5 million times more energy than the equivalent weight of coal. Uranium has also been used as a weapon, in the atomic bomb that was detonated over Hiroshima on 6 August 1945. This bomb, which contained 64kg (141lb) of enriched uranium, was responsible for the deaths of between 70,000 and 126,000 civilians. Today, depleted uranium, which is 68.4 per cent denser than lead, is used in armour-piercing weaponry and armour plating.

# Neptunium

DISCOVERY DATE:1940    DISCOVERED BY: Edwin McMillan and Philip H. Abelson

Neptunium is a silver-coloured metal that tarnishes when exposed to air. It melts at around 640°C and its boiling point has been calculated to be 4174°C (7545.2°F), although this latter number has never been verified. This may give it the title of the element with the greatest temperature range in a liquid state, which would otherwise belong to gallium. Like thorium, neptunium will spontaneously ignite in air when it is divided into small grains.

Neptunium is one of only two transuranium elements – that is, the elements that are heavier than uranium, element 92 – to occur in nature, the other being plutonium. Both of these elements are only found in trace amounts. It was discovered in 1940 at the Berkeley Radiation Laboratory of the University of California, Berkeley by Edwin McMillan and Philip Abelson, who bombarded atoms of uranium-238 with neutrons

Trace amounts of neptunium-237 and neptunium-239 can be found in uraninite, a uranium-rich mineral and ore.

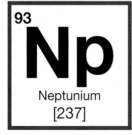

93
**Np**
Neptunium
[237]

| Atomic number: | 93 |
| Group: | Actinides |
| Period: | Period 7 |
| Block: | f-block |
| Atomic mass: | [237] |
| Melting point: | 639±3 °C, 1182±5 °F |
| Boiling point: | 4174 °C, 7545 °F (extrapolated) |
| Density: | 20.45 g/cm³ (alpha, near room temperature) |
| Appearance: | Silvery metal |

**Np**

Edwin McMillan (1907–1991), was an American physicist who was awarded the Nobel Prize in Chemistry with Glenn Seaborg in 1951 for their discovery of neptunium.

in a 1.52-m (60-in) cyclotron and subsequently demonstrated that they had produced neptunium-239, an isotope with a half-life of 2.3 days. The new element was named after the planet Neptune to reflect its position in the periodic table next to uranium, which had been named after Neptune's neighbouring planet, Uranus.

There are few industrial applications of neptunium, although in principle it could be used as fuel for a nuclear reactor or as a weapon. Approximately 2500kg (5511lb) of neptunium were released into the atmosphere during nuclear testing (until the partial nuclear Test Ban Treaty of 1963) and this accounts for most of the neptunium now in the environment. Like protactinium, tiny amounts of neptunium are generated in our homes through the radioactive decay of americium-241 in domestic smoke detectors.

# Plutonium

DISCOVERY DATE: 1940–41 DISCOVERED BY: Glenn T. Seaborg, Arthur Wahl, Joseph W. Kennedy & Edwin McMillan

Plutonium is a radioactive silvery-white metal in the actinide series that forms a dull grey oxidized coating when exposed to the air. It was first produced by a team at Berkeley, California, in December 1940, who bombarded atoms of uranium-238 with deuterons (the nuclei of deuterium atoms, each containing one proton and one neutron) in a cyclotron. This process created atoms of neptunium-238, which subsequently decayed into plutonium-238.

Due to wartime secrecy, the discovery of the new element 94 was kept secret until 1948. As the next element in line after uranium and neptunium, plutonium was named after Pluto, which at that time was considered to be a planet. (Pluto was subsequently demoted to the status of dwarf planet in 2006 by the International Astronomical Union.)

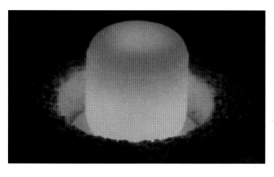

This pellet of plutonium-238, glowing with its own light, is an ideal power source for space probes, satellites and scientific instruments left on the Moon by Apollo astronauts.

| 94 | |
|---|---|
| **Pu** | |
| Plutonium | |
| [244] | |

| | |
|---|---|
| **Atomic number:** | 94 |
| **Group:** | Actinides |
| **Period:** | Period 7 |
| **Block:** | f-block |
| **Atomic mass:** | [244] |
| **Melting point:** | 639.4 °C, 1182.9 °F |
| **Boiling point:** | 3228 °C, 5842 °F |
| **Density:** | 19.85 g/cm³ (near room temperature) |
| **Appearance:** | Silvery-white metal |

An artist's impression of Pluto, the dwarf planet that gave its name to plutonium.

Plutonium is the heaviest element known to occur naturally, being created when natural deposits of uranium-238 capture neutrons released by the decay of other uranium-238 atoms. The isotope plutonium-244 has a half-life of 80.8 million years, which means that it should be possible to trace tiny amounts that were present when the Earth was formed 4.5 billion years ago, but so far none have been detected.

Plutonium was used in the atomic bomb detonated over Nagasaki on 9 August 1945. The 6.4kg (14.1lb) of plutonium contained in the bomb had the same explosive yield that would be obtained from 20 kilotons (22,046 tons) of TNT, and resulted in the deaths of 60,000–80,000 people. Plutonium is still used in today's nuclear weapons. Another use for plutonium is as a source of heat and power for space probes and Mars rovers.

# Americium

DISCOVERY DATE: 1944 DISCOVERED BY: Glenn T. Seaborg, Ralph A. James, Leon O. Morgan & Albert Ghiorso

Americium is a soft silvery-white metal in the actinide series that tarnishes gradually when exposed to air. It is radioactive and has approximately 19 isotopes and 11 nuclear isomers. It was first produced in 1944 as part of the Manhattan Project, led by Glenn T. Seaborg, and its discovery was kept secret until November 1945. Although it might seem to be named after a single country, Americium in fact took its name from the American continents, to mark its location directly below europium in the f-block of the periodic table.

Like all the radioactive elements, americium is dangerous to human health and should be handled with extreme care. It is surprising, therefore, that americium is present in millions of homes and other buildings, where it provides a life-saving

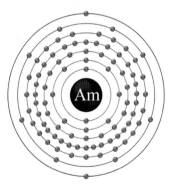

This diagram shows the atomic structure of an americium-243 atom, with 95 electrons orbiting the nucleus containing 95 protons and 148 neutrons.

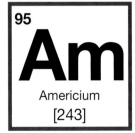

**95**

**Am**

Americium

[243]

| | |
|---|---|
| **Atomic number:** | 95 |
| **Group:** | Actinides |
| **Period:** | Period 7 |
| **Block:** | f-block |
| **Atomic mass:** | [243] |
| **Melting point:** | 1176 °C, 2149 °F |
| **Boiling point:** | 2607 °C, 4725 °F (calculated) |
| **Density:** | 12 g/cm³ (near room temperature) |
| **Appearance:** | Soft silvery-white metal |

Americium may be extremely rare and precious, but many of us have tiny amounts of it protecting our lives in our domestic smoke detection equipment.

function. This is because americium-241 is the active substance that reacts to the presence of smoke in smoke detectors. The radiation from a tiny piece of americium foil ionises the electrons between two metal plates, allowing an electric current to pass between the plates. If any smoke particles enter the ionisation chamber, the current is interrupted, and this triggers the detector's alarm. These detectors can identify smoke particles that would be missed by an optical smoke detector, and the amount of radiation emitted in normal use is lower than natural background radiation levels. Even if a person opened the sealed chamber of a smoke detector and swallowed or inhaled the americium inside, the radiation would still only match background radiation levels – but such a course of action is not recommended.

# Curium

DISCOVERY DATE: 1944  DISCOVERED BY: Glenn T. Seaborg, Ralph A. James, Albert Ghiorso

Curium is a silver-coloured metal in the actinide series that is hard, dense and so radioactive that it emits a purple glow. It was first produced in 1944 by Glenn T. Seaborg, Ralph James and Albert Ghiorso, by using the cyclotron at Berkeley, California to bombard plutonium-239 with alpha particles. The resulting sample was then analyzed by researchers at the University of Chicago, who were able to chemically identify is as the new element 96.

Curium was named in honour of Marie Skłodowska-Curie and Pierre Curie, in recognition of their discovery of radium and their research into radioactivity. This decision continued the pattern set by lanthanide europium and actinide americium, because curium lies directly below the lanthanide gadolinium in the periodic table. Like curium, gadolinium was also named in honour of a scientist, in this case the Finnish chemist, physicist and mineralogist Johan Gadolin. In a break

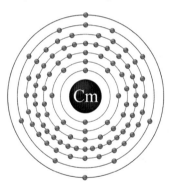

The atomic structure of a curium-247 atom is made up of a nucleus containing 96 protons and 151 neutrons, orbited by 96 electrons.

| 96 | |
|---|---|
| **Cm** | |
| Curium | |
| [247] | |

| | |
|---|---|
| Atomic number: | 96 |
| Group: | Actinides |
| Period: | Period 7 |
| Block: | f-block |
| Atomic mass: | [247] |
| Melting point: | 1340 °C, 2444 °F |
| Boiling point: | 3110 °C, 5630 °F |
| Density: | 13.51 g/cm³ (near room temperature) |
| Appearance: | Silvery metal |

Pierre Curie (1859–1906) and Marie Skłodowska-Curie (1867–1934) were pioneers in the fields of radioactivity, crystallography and magnetism. Their combined achievements are recognised in the name of the element curium, which was discovered in 1944.

with convention, the discovery of curium was first announced on a children's radio programme called *Quiz Kids* in November 1945, five days before its official news release at a meeting of the American Chemical Society.

There are around 19 known isotopes and 7 nuclear isomers of curium, none of which are stable. Curium-247 has the longest half-life, at 15.6 million years. The isotopes that are used most frequently are curium-242, with a half-life of 162.8 days, and curium-244, which has a half-life of 18.1 years. Curium-244 can be used as a source of alpha particles, which makes it useful in the X-ray spectrometers on space probes. It is also being studied as a potential power source for radioisotope thermoelectric generators in spacecraft.

Actinides

# Berkelium

DISCOVERY DATE: 1949 DISCOVERED BY: Glenn T. Seaborg, Albert Ghiorso, Stanley G. Thompson & Kenneth Street Jr.

Berkelium takes its name from the city of Berkeley, California, where it was first created in December 1949 by a team led by Glenn T. Seaborg at Lawrence Berkeley National Laboratory. It was produced by bombarding americium-241 with helium nuclei in a 152-cm (60-in) cyclotron for several hours.

Since 1967, just over 1g (0.035oz) of berkelium has been produced in the United States. Due to its intense radioactivity, berkelium would be highly dangerous to human health, but so little has ever been created that this is essentially a theoretical risk.

Several isotopes of berkelium have been described, and all of them are radioactive. Berkelium-247 has the longest half-life, at 1380 years, followed by berkelium-248 (more than 300 years) and berkelium-249 (330 days).

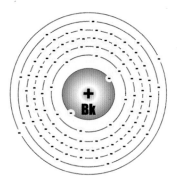

This diagram shows the atomic structure of berkelium, with mass and energy levels.

**97**

# Bk

Berkelium

[247]

| | |
|---|---|
| **Atomic number:** | 97 |
| **Group:** | Actinides |
| **Period:** | Period 7 |
| **Block:** | f-block |
| **Atomic mass:** | [247] |
| **Melting point:** | 986 °C, 1807 °F (beta) |
| **Boiling point:** | 2627 °C, 4760 °F (beta) |
| **Density:** | 14.78 g/cm3 (alpha, near room temperature), 13.25 g/cm3 (beta, near room temperature) |
| **Appearance:** | Soft silvery-white metal |

# Californium

**Cf**

DISCOVERY DATE: 1950 DISCOVERED BY: Glenn T. Seaborg, Albert Ghiorso, Stanley G. Thompson and Kenneth Street Jr.

Californium, named after the university (Lawrence Berkeley National Laboratory) and state in which it was discovered, was created by bombarding curium-242 with helium-4 ions. Although the researchers had needed three years to create enough curium to produce the new element, the isotope created through this process, californium-245, has a half-life of only 44 minutes.

As a highly radioactive element, californium is very harmful to human bodies if breathed in or absorbed through food or drink. However, it does have medical applications: it has been used successfully as a treatment for cervical and brain cancer, with a tiny amount of the element being transplanted directly into the tumour. Californium is also used for industrial purposes, including as a neutron source for nuclear reactors and in portable metal detectors.

Californium-252 has a number of specialized uses as a strong neutron emitter; it produces 139 million neutrons per microgram per minute.

| 98 | | |
|---|---|---|
| **Cf** | | |
| Californium | | |
| [251] | | |

| | |
|---|---|
| **Atomic number:** | 98 |
| **Group:** | Actinides |
| **Period:** | Period 7 |
| **Block:** | f-block |
| **Atomic mass:** | [251] |
| **Melting point:** | 900 °C, 1652 °F |
| **Boiling point:** | 1470 °C, 2678 °F (estimated) |
| **Density:** | 15.1 g/cm³ (near room temperature) |
| **Appearance:** | Soft silvery-white metal |

# Einsteinium

DISCOVERY DATE: 1952 DISCOVERED BY: Albert Ghiorso and colleagues at Lawrence Berkeley National Laboratory

Einsteinium is a soft, silvery, highly radioactive metal, named after the famous theoretical physicist Albert Einstein, who developed the general theory of relativity and won the 1921 Nobel Prize in Physics for his discovery of the law of the photoelectric effect.

Einsteinium was discovered in 1952, in the debris of the first hydrogen bomb test, which took place on the Enewetak Atoll in the Pacific Ocean. This test was carried out by the United States, with the code name Ivy Mike. It had 700 times the energy of the atomic bomb attack on Hiroshima, and completely destroyed a small island called Elugelab.

Due to the intense secrecy around the test, the discovery was not made public until 1955, and it was not until 1961 that enough of the element was created for it to be visible to the naked eye. Even today, only around 1 milligram of einsteinium is created each year.

Einsteinium was named after Albert Einstein (1879–1955), one of the most influential scientists of all time. He is best known for developing the theory of relativity and for his contributions to the field of quantum mechanics.

## 99 Es Einsteinium [252]

| | |
|---|---|
| Atomic number: | 99 |
| Group: | Actinides |
| Period: | Period 7 |
| Block: | f-block |
| Atomic mass: | [252] |
| Melting point: | 860 °C, 1580 °F |
| Boiling point: | 996 °C, 1825 °F (estimated) |
| Density: | 8.84 g/cm³ (near room temperature) |
| Appearance: | Soft silvery metal |

# Fermium

**Fm**

DISCOVERY DATE: 1952 DISCOVERED BY: Albert Ghiorso and colleagues at Lawrence Berkeley National Laboratory

Fermium was named after Enrico Fermi, an Italian-American physicist who was awarded the 1938 Nobel Prize in Physics for his work on radioactivity. It is the heaviest element that can be produced by bombarding lighter elements with neutrons. Like einsteinium, fermium was discovered in the fallout from the Ivy Mike hydrogen bomb test.

All the elements from number 100 upwards are called the 'transfermium' elements – meaning the elements after fermium. All of these elements have high levels of radioactivity, and because of this, none of them are found naturally occurring on Earth. This does not mean that they never existed on Earth – it's just that they are so unstable that they would have decayed a very long time ago.

The Ivy Mike hydrogen bomb test in 1952 led to the discovery of two new elements: einsteinium and fermium.

| 100 |
| --- |
| **Fm** |
| Fermium |
| [257] |

| Atomic number: | 100 |
| --- | --- |
| Group: | Actinides |
| Period: | Period 7 |
| Block: | f-block |
| Atomic mass: | [257] |
| Melting point: | 500 °C, 2800 °F (predicted) |
| Boiling point: | Unknown |
| Density: | 9.71 g/cm³ (near room temperature, predicted) |
| Appearance: | Unknown |

# Md

# Mendelevium

DISCOVERY DATE: 1955     DISCOVERED BY: Albert Ghiorso and colleagues

Element 101 is named after the father of the periodic table, Dmitri Mendeleev. Although he could almost certainly not have predicted how far his table of elements would extend – and even today's scientists are open to the idea that there are many more elements still to be discovered – what is truly remarkable about Mendeleev's work is that he predicted the existence of as-yet-undiscovered elements, based on his analysis of the properties of the currently known ones.

Mendelevium is created by bombarding einsteinium atoms with alpha particles. Like all the elements at this end of the table, mendelevium is only used for research purposes and has no commercial applications.

M. Stanley Livingston (left) and Ernest O. Lawrence (right) standing in front of the 27-inch cyclotron at the old Radiation Laboratory at the University of California, Berkeley.

| 101 | |
|---|---|
| **Md** | |
| Mendelevium | |
| [258] | |

| | |
|---|---|
| Atomic number: | 101 |
| Group: | Actinides |
| Period: | Period 7 |
| Block: | f-block |
| Atomic mass: | [258] |
| Melting point: | 800 °C, 1500 °F (predicted) |
| Boiling point: | Unknown |
| Density: | 10.37 g/cm³ (near room temperature, predicted) |
| Appearance: | Unknown |

# Nobelium

**No**

DISCOVERY DATE: 1963 DISCOVERED BY: Researchers at the Joint Institute for Nuclear Research, Dubna, Russia

Nobelium, the second-last member of the actinide series, is a radioactive metal which is produced by bombarding curium with carbon in a cyclotron. It has 12 known isotopes, of which nobelium-259 has the longest half-life, at 58 minutes. Nobelium-255 has a much shorter half-life, at just over 3 minutes, but as it can be produced more easily than other isotopes, it is most commonly used in chemistry research.

Nobelium was named after Alfred Nobel, the inventor of dynamite, who left his fortune after his death to establish the Nobel Prizes. This name was chosen by Swedish scientists who mistakenly thought they had discovered element 102 in 1957. It was then synthesized independently by teams in Dubna, Russia and Berkeley, California, and in 1997 IUPAC credited the Russian team as the official discoverers.

Alfred Nobel (1833–1896) was a scientist and businessman most famed in his lifetime for the invention of dynamite. In his will, he left his fortune to establish six prizes for remarkable achievements in physics, chemistry, physiology or medicine, literature, and peace.

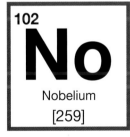

102
## No
Nobelium
[259]

| Atomic number: | 102 |
|---|---|
| Group: | Actinides |
| Period: | Period 7 |
| Block: | f-block |
| Atomic mass: | [259] |

| Melting point: | 800 °C, 1500 °F (predicted) |
|---|---|
| Boiling point: | Unknown |
| Density: | 9.94 g/cm³ (near room temperature, predicted) |
| Appearance: | Unknown |

# Lr
# Lawrencium

DISCOVERY DATE: 1961–71      DISCOVERED BY: Researchers in the USA & Russia

Lawrencium is the final element in the actinide group, and was named after the nuclear physicist Ernest Lawrence. Lawrence was awarded the Nobel Prize in Physics in 1939 for his invention of the cyclotron, a type of particle accelerator that was essential in the discovery of these radioactive elements at the far end of the periodic table.

Lawrencium is a radioactive metal with 14 known isotopes. It was first produced at Lawrence Berkeley National Laboratory by bombarding three isotopes of californium with nuclei of boron-10 and boron-11. Researchers at the Joint Institute for Nuclear Research in Dubna, Russia (then the Soviet Union), also claimed the discovery, and in 1992 the IUPAC recognized both teams as co-discoverers.

In 1929, Ernest Lawrence (1901–1958) became intrigued by the idea of a device that could produce high-energy particles, and he went on to invent the cyclotron, with its circular accelerating chamber.

**103**
# Lr
Lawrencium
[262]

| Atomic number: | 103 |
| --- | --- |
| Group: | Actinides |
| Period: | Period 7 |
| Block: | f-block |
| Atomic mass: | [262] |

| Melting point: | 1600 °C, 3000 °F (predicted) |
| --- | --- |
| Boiling point: | Unknown |
| Density: | 14.4 g/cm³ (near room temperature, predicted) |
| Appearance: | Unknown |

# Rutherfordium

Rf

DISCOVERY DATE: 1969     DISCOVERED BY: Researchers in the USA & Russia

Rutherfordium is a radioactive element of which only a small number of atoms have ever been produced. It is created by bombarding californium atoms with carbon nuclei.

This element was named after the British physicist Ernest Rutherford, who received the Nobel Prize in Chemistry in 1908 for his work on the chemistry of radioactive substances. Rutherford developed the orbital theory of the atom and has been described as 'the father of modern physics'. As with the previous element, Lawrencium, researchers in both Dubna, Russia and Berkeley, California were attempting to synthesize element 104 in the 1960s, and both teams claimed the discovery as their own. In 1997, the IUPAC officially named both sets of researchers as co-discoverers.

New Zealand physicist Ernest Rutherford (1871–1937) has been described as 'the greatest experimentalist since Michael Faraday' and became the first Nobel laureate from Oceania.

| 104 | |
|---|---|
| **Rf** | |
| Rutherfordium | |
| [267] | |

| | |
|---|---|
| Atomic number: | 104 |
| Group: | Group 4 |
| Period: | Period 7 |
| Block: | d-block |
| Atomic mass: | [267] |
| Melting point: | 2100 °C, 3800 °F (predicted) |
| Boiling point: | 5500 °C, 9900 °F (predicted) |
| Density: | 17 g/cm3 (near room temperature, predicted) |
| Appearance: | Unknown |

# Db

# Dubnium

DISCOVERY DATE: 1970     DISCOVERED BY: Researchers in the USA & Russia

Dubnium is a highly radioactive synthetic chemical element, and only a few atoms of it have ever been made. Its most stable isotope, dubnium-268, has a half-life of 28 hours.

In 1967, researchers at the Joint Institute for Nuclear Research in Dubna, Russia, created what they believed to be element 105 by bombarding atoms of americium-243 with a beam of neon-22 ions, and they were able to confirm their findings in 1970.

Also in 1970, a team at the Lawrence Berkeley Laboratory in California synthesized the same element, by bombarding californium-249 with nitrogen-15 nuclei. Both teams were recognized as co-discoverers and the name of the new element was eventually finalized as dubnium in honour of Dubna in 1997.

American nuclear scientist Albert Ghiorso (1915–2010) was a co-discoverer of twelve elements in the periodic table, a feat that is unmatched by any other scientist to date.

## 105

# Db

Dubnium

[262]

| Atomic number: | 105 |
| --- | --- |
| Group: | Group 5 |
| Period: | Period 7 |
| Block: | d-block |
| Atomic mass: | [262] |

| Melting point: | Unknown |
| --- | --- |
| Boiling point: | Unknown |
| Density: | 21.6 g/cm³ (near room temperature, predicted) |
| Appearance: | Unknown |

# Seaborgium

**Sg**

DISCOVERY DATE: 1974    DISCOVERED BY: Researchers at Lawrence Berkeley National Laboratory, USA

Seaborgium was named in honour of American nuclear chemist Glenn T. Seaborg, who was awarded the 1951 Nobel Prize in Chemistry for his work on the synthesis and chemistry of ten of the transuranium elements. It is the first element to have been named after someone who was still alive.

The most stable isotope of seaborgium has a half-life of 14 minutes, which means it is only useful for research purposes. Although too few atoms have ever been produced to confirm this, it is predicted to be a solid under normal conditions and to share some structural properties with tungsten, the element directly above it in the periodic table.

Glenn T. Seaborg (1912–1999) was an American nuclear chemist who was involved in the synthesis and discovery of ten of the transuranium elements.

| 106 |
| --- |
| **Sg** |
| Seaborgium |
| [269] |

| | |
| --- | --- |
| **Atomic number:** | 106 |
| **Group:** | Group 6 |
| **Period:** | Period 7 |
| **Block:** | d-block |
| **Atomic mass:** | [269] |
| **Melting point:** | Unknown |
| **Boiling point:** | Unknown |
| **Density:** | 23–24 g/cm³ (near room temperature, predicted) |
| **Appearance:** | Unknown |

**Bh**

# Bohrium

DISCOVERY DATE: 1981 DISCOVERED BY: Researchers at the GSI Helmholtz Centre for Heavy Ion Research, Germany

Bohrium is a highly radioactive element whose most stable isotope, bohrium-270, has a half-life of 2.4 minutes. It was created by a team of German researchers, and named after the Danish scientist Niels Bohr, who won the Nobel Prize in Physics in 1922 for his work on atomic structure and quantum theory.

The original name proposed by the team for this element was nielsbohrium, as the researchers were worried about the potential for the new name to be confused with boron, but this was rejected by the International Union of Pure and Applied Chemistry becausethere was no precedent for including a person's first name in the name of an element.

Danish physicist Niels Bohr (1885–1862) contributed to our understanding of quantum theory and atomic structure. In the 1930s he helped refugees from Germany to find scientific work in other countries, and in 1940 he donated his Nobel Prize medal to raise money for the Finnish Relief Fund.

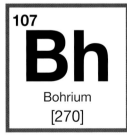

**107**

**Bh**

Bohrium

[270]

| | |
|---|---|
| **Atomic number:** | 107 |
| **Group:** | Group 7 |
| **Period:** | Period 7 |
| **Block:** | d-block |
| **Atomic mass:** | [270] |

| | |
|---|---|
| **Melting point:** | Unknown |
| **Boiling point:** | Unknown |
| **Density:** | 26–27 g/cm³ (near room temperature, predicted) |
| **Appearance:** | Unknown |

Transition metals

# Hassium

**Hs**

DISCOVERY DATE: 1984    DISCOVERED BY: Researchers at the GSI Helmholtz Centre, Germany

Hassium is a superheavy element that is highly radioactive and decays rapidly: the most stable of its known isotopes have half-lives of approximately 10 seconds. It was discovered by a team led by Peter Armbruster and Gottfried Münzenberg in Darmstadt, Germany, and named after the German state of Hesse, which is 'Hassia' in Latin, where it was first created.

Although only around 100 atoms of hassium have been created since it was first synthesized in 1984, it is predicted to share some of the properties of osmium, the element directly above it in Group 8, and to be one of the densest elements in the entire periodic table.

The GSI Helmholtzzentrum für Schwerionenforschung in Darmstadt is a research facility that was founded in 1969 to conduct research on and with heavy ion accelerators. It is the only large-scale research facility in Hesse.

| 108 |
| --- |
| **Hs** |
| Hassium |
| [269] |

| | |
| --- | --- |
| **Atomic number:** | 108 |
| **Group:** | Group 8 |
| **Period:** | Period 7 |
| **Block:** | d-block |
| **Atomic mass:** | [269] |
| **Melting point:** | Unknown |
| **Boiling point:** | Unknown |
| **Density:** | 27–29 g/cm³ (near room temperature, predicted) |
| **Appearance:** | Unknown |

**Mt**

# Meitnerium

DISCOVERY DATE: 1982     DISCOVERED BY: Researchers at the GSI Helmholtz Centre, Germany

Meitnerium is a highly radioactive element, discovered in 1982 in Darmstadt, Germany. The most stable of its known isotopes, meitnerium-278, has a half-life of only 4.5 seconds, although it is possible that meitnerium-282 may have a longer half-life of just over a minute.

Meitnerium has the unusual distinction as the only element to have been named after an individual non-mythological woman. (Although Marie Skłodowska-Curie's contributions to science were recognized in the naming of curium, this honour was shared with her husband Pierre and did not belong to her alone.) Lise Meitner was an Austrian-Swedish physicist who was responsible for the co-discovery of the element protactinium and the process of nuclear fission.

At a ceremony at the GSI in 1992, the official names for elements 107, 108 and 109 were confirmed: Nielsbohrium (which was later changed to Bohrium), Hassium and Meitnerium respectively.

| | |
|---|---|
| **109** | |
| **Mt** | |
| Meitnerium | |
| [278] | |

| | |
|---|---|
| Atomic number: | 109 |
| Group: | Group 9 |
| Period: | Period 7 |
| Block: | d-block |
| Atomic mass: | [278] |

| | |
|---|---|
| Melting point: | Unknown |
| Boiling point: | Unknown |
| Density: | 27–28 g/cm³ (near room temperature, predicted) |
| Appearance: | Unknown |

# Darmstadtium

**Ds**

DISCOVERY DATE: 1994     DISCOVERED BY: Researchers at the GSI Helmholtz Centre, Germany

Darmstadtium was discovered in 1994 at the GSI Helmholtz Centre for Heavy Ion Research in Darmstadt, Germany. It is so unstable that only individual atoms have ever been synthesized. As a result, not much is known about its properties, but it is predicted to be a metal that is solid at room temperature and that resists corrosion and oxidation.

Darmstadtium has 15 known isotopes, the most stable of which has a half-life of 4 minutes. Alternative names proposed for this element were wixhausium, in honour of the suburb in which the lab is located, and, jokingly, policium, to mark the fact that the emergency telephone number for the police service in Germany is 110 – but darmstadtium is the name that was made official in 2003 by the IUPAC.

Sigurd Hofmann, head of the Heavy Elements Program at the GSI, led the discovery experiments of the elements darmstadtium, roentgenium and copernicium.

| | |
|---|---|
| **110** | |
| **Ds** | |
| Darmstadtium | |
| [281] | |

| | |
|---|---|
| Atomic number: | 110 |
| Group: | Group 10 |
| Period: | Period 7 |
| Block: | d-block |
| Atomic mass: | [281] |
| Melting point: | Unknown |
| Boiling point: | Unknown |
| Density: | 26–27 g/cm$^3$ (near room temperature, predicted) |
| Appearance: | Unknown |

# Roentgenium

DISCOVERY DATE: 1994    DISCOVERED BY: Researchers at the GSI Helmholtz Centre, Germany

Like many of the elements at this far end of the periodic table, roentgenium had a placeholder name before it was formally discovered. As element 111, it had a Latin name spelling out its number, which turned out to be the rather pleasing 'unununium', with chemical symbol Uuu.

Researchers at the GSI Helmholtz Centre for Heavy Ion Research were able to bring this numerical alias to an end when they bombarded bismuth-209 with nickel-64 and succeeded in synthesizing an atom of roentgenium-272. The new element was named to honour the physicist Wilhelm Roentgen, who won the first Nobel Prize in Physics in 1901 for his discovery of X-rays.

**Wilhelm Conrad Roentgen (1845–1923) was a German chemist and mechanical engineer who discovered how to produce and detect X-rays in 1895. Element 111 was named roentgenium in his honour.**

**111**

# Rg

Roentgenium
[282]

| Atomic number: | 111 |
| --- | --- |
| Group: | Group 11 |
| Period: | Period 7 |
| Block: | d-block |
| Atomic mass: | [282] |

| Melting point: | Unknown |
| --- | --- |
| Boiling point: | Unknown |
| Density: | 22–24 g/cm³ (near room temperature, predicted) |
| Appearance: | Unknown |

# Copernicium

**Cn**

DISCOVERY DATE: 1996    DISCOVERED BY: Researchers at the GSI Helmholtz Centre, Germany

Copernicium is a highly radioactive element in Group 12 of the periodic table, directly below mercury. It was synthesized by researchers in Darmstadt, Germany, who bombarded atoms of lead-208 with zinc-70 nuclei to produce one atom of the new element. The researchers suggested the name copernicium to honour Nicolaus Copernicus, the mathematician and astronomer who proposed a heliocentric model of the Universe.

Not enough copernicium has ever been produced for its physical characteristics to be fully understood, but it is predicted to be a silver transition metal that is liquid at room temperature but behaves like a noble gas. Seven isotopes of copernicium have been confirmed to date, the most stable of which has a half-life of around 30 seconds.

**Copernicium was synthesised in 1996 by an international team of scientists at GSI Helmholtz Centre for Heavy Ion Research in Darmstadt, Germany.**

**112**

# Cn

Copernicium

[285]

| Atomic number: | 112 |
| --- | --- |
| Group: | Group 12 |
| Period: | Period 7 |
| Block: | d-block |
| Atomic mass: | [285] |

| Melting point: | 10±11 °C, 50±20 °F (predicted) |
| --- | --- |
| Boiling point: | 67±10 °C, 153±18 °F (predicted) |
| Density: | 14.0 g/cm³ (near room temperature, predicted) |
| Appearance: | Unknown |

Until the middle of 2016, the unnamed element filling space 113 on the periodic table had the somewhat uninspiring name of 'ununtrium', meaning 'one one three' in Latin, but all that changed in June of that year, when the International Union of Pure and Applied Chemistry announced the official names of elements 113, 115, 117 and 118. The names were formally approved in November 2016.

Nihonium, an extremely radioactive element, was discovered by a research team at RIKEN, the Institute of Physical and Chemical Research in Wakō, Japan. It was the first element to be discovered in an Asian country, and its name comes from 'Nihon', the Japanese word for Japan.

A proud moment: Kosuke Morita, a researcher at RIKEN, points to element 113 on a periodic table of the elements at a press conference on 9 June 2016, to mark its official naming as nihonium, with chemical symbol Nh.

### 113
# Nh
### Nihonium
### [286]

| | |
|---|---|
| Atomic number: | 113 |
| Group: | Group 13 |
| Period: | Period 7 |
| Block: | p-block |
| Atomic mass: | [286] |
| Melting point: | 430 °C, 810 °F (predicted) |
| Boiling point: | 1130 °C, 2070 °F (predicted) |
| Density: | 16.0 g/cm³ (near room temperature, predicted) |
| Appearance: | Unknown |

# Flerovium

**FI**

DISCOVERY DATE: 1999    DISCOVERED BY: Researchers from Russia and USA

Flerovium is a highly radioactive superheavy element that was discovered in Russia in 1999. The new element took its name from the Flerov Laboratory of Nuclear Reactions at the Joint Institute for Nuclear Research, and the laboratory itself was named in honour of the Russian physicist Georgy Flyorov.

Flerovium has six known isotopes, with atomic weights ranging from flerovium-284 to flerovium-289. The last of these is the most stable, and even this has a half-life of just under 2 seconds. Experiments have suggested that it is a very volatile element, and may be a gas at room temperature, although it is still unclear whether flerovium more closely resembles a metal or a noble gas.

**Russian physicist Georgy Flyorov (1913–1990) is known for his work in crystallography and materials science. Element 114 was named flerovium in his honour in 2012.**

| 114 |
|---|
| **FI** |
| Flerovium |
| [289] |

| Atomic number: | 114 |
|---|---|
| Group: | Group 14 |
| Period: | Period 7 |
| Block: | p-block |
| Atomic mass: | [289] |

| Melting point: | 11±50 °C, 52±90 °F (predicted) |
|---|---|
| Boiling point: | Unknown |
| Density: | 11.4±0.3 g/cm3 (near room temperature, predicted) |
| Appearance: | Expected to be silvery, metallic and solid at room temperature |

# Moscovium

DISCOVERY DATE: 2003     DISCOVERED BY: Researchers from Russia and USA

Moscovium, a highly radioactive and unstable element, was synthesized in 2003 through a collaboration between teams at the Joint Institute for Nuclear Research in Russia, led by Yuri Oganessian, and the Lawrence Livermore National Laboratory in the USA, led by Ken Moody. The researchers bombarded americium-243 with calcium-48 nuclei ions, and produced four atoms of the new element. These atoms decayed into nihonium atoms in about 100 milliseconds.

The discovery of element 115 was officially confirmed in December 2015, and it was named moscovium in 2016 after the Moscow Oblast region, where the Joint Institute for Nuclear Research is based. To date, over 100 atoms of moscovium have been created, with mass numbers ranging from 296 to 290.

A view of the main block housing the proton synchrotron at the Joint Nuclear Research Institute in the Soviet Union, in 1956.

## 115 Mc
Moscovium
[289]

| | |
|---|---|
| Atomic number: | 115 |
| Group: | Group 15 |
| Period: | Period 7 |
| Block: | p-block |
| Atomic mass: | [289] |

| | |
|---|---|
| Melting point: | 400 °C, 750 °F (predicted) |
| Boiling point: | ~1100 °C, ~2000 °F (predicted) |
| Density: | 13.5 g/cm³ (near room temperature, predicted) |
| Appearance: | Unknown |

# Livermorium

**Lv**

DISCOVERY DATE: 2000     DISCOVERED BY: Researchers from Russia and USA

The previous two elements in the periodic table, flerovium and moscovium, were both named after Russia, and names of elements 116 and 117 take us back to the USA. Livermorium is named after the Lawrence Livermore National Laboratory in the city of Livermore, California, where much of the research that led to the discovery of this element took place.

Like all the elements in this section of the table, livermorium is a highly radioactive element, and not enough of it has ever been produced for us to know much about its characteristics and likely behaviour. Five isotopes of this element are known, the longest-lived of which, livermorium-293, has a half-life of around 60 milliseconds.

An aerial view of Lawrence Livermore National Laboratory in Livermore, California. This institution was originally established as the University of California Radiation Laboratory in 1952.

**116**

**Lv**

Livermorium

[293]

| | |
|---|---|
| **Atomic number:** | 116 |
| **Group:** | Group 16 |
| **Period:** | Period 7 |
| **Block:** | p-block |
| **Atomic mass:** | [293] |
| **Melting point:** | 364–507 °C, 687–944 °F (extrapolated) |
| **Boiling point:** | 762–862 °C, 1403–1583 °F (extrapolated) |
| **Density:** | 12.9 g/cm3 (near room temperature, predicted) |
| **Appearance:** | Unknown |

# Tennessine

DISCOVERY DATE: 2009    DISCOVERED BY: Researchers from Russia and the USA

Tennessine, like its neighbour livermorium, was produced by a collaboration between scientists in Russia and the USA, marking a welcome and positive change after the controversies around discovery and naming that hung over many of the earlier transfermium elements. Sadly, such collaborations came to an end in 2022 following Russia's invasion of Ukraine.

Tennessine was synthesized by bombarding atoms of berkelium-249 with a beam of calcium-48, resulting in the creation of six atoms of the new element. To date, very little is known about the properties of tennessine, because it is so difficult and expensive to produce, and because it decays extremely quickly due to its radioactivity. The most stable isotope of tennessine has a half-life of just 51 milliseconds.

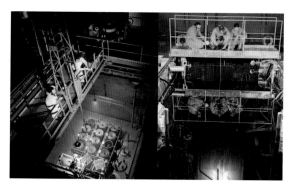

Isotope reactor and fuel pool at Oak Ridge National Laboratory, the element is named after Tennessee in recognition of Tennessee's contributions to its discovery.

## 117

## Ts

Tennessine

[294]

| | |
|---|---|
| Atomic number: | 117 |
| Group: | Group 17 |
| Period: | Period 7 |
| Block: | p-block |
| Atomic mass: | [294] |

| | |
|---|---|
| Melting point: | 350–550 °C, 662–1022 °F (predicted) |
| Boiling point: | 610 °C, 1130 °F (predicted) |
| Density: | 7.1–7.3 g/cm³ (near room temperature, extrapolated) |
| Appearance: | Unknown |

# Oganesson

**Og**

DISCOVERY DATE: 2002    DISCOVERED BY: Researchers from Russia and the USA

With oganesson, we come to the last known element in the periodic table, and the only one to be named after someone who is still alive at the time of writing. Yuri Oganessian, who was born in 1933, is a Russian nuclear physicist who played a key role in the discovery of several elements. His invention of the 'cold fusion' method was fundamental to the synthesis of elements 107 to 113, and his later 'hot fusion' method aided the synthesis of elements 114 to 118.

From its position in Group 18 of the table, oganesson should in theory display the characteristics of a noble gas. However, research on the element has suggested it may be reactive and solid at room temperature. By 2020, only five atoms of oganesson had been created, and more research will be needed before we can understand how this unusual element fits into Dmitri Mendeleev's history-making periodic table.

This Armenian stamp, issued in 2017, celebrates the achievements of Russian–Armenian nuclear physicist Yuri Oganessian, who played a key role in the discovery of the superheavy elements.

| | |
|---|---|
| **Atomic number:** | 118 |
| **Group:** | Group 18 |
| **Period:** | Period 7 |
| **Block:** | p-block |
| **Atomic mass:** | [294] |

| | |
|---|---|
| **Melting point:** | 52±15 °C, 125±27 °F (predicted) |
| **Boiling point:** | 177±10 °C, 350±18 °F (predicted) |
| **Density:** | 7.2 g/cm3 (near room temperature) |
| **Appearance:** | Unknown |

# Index of elements (listed alphabetically)

Actinium...186
Aluminium...34
Americium...198
Antimony...110
Argon...44
Arsenic...74
Astatine...178
Barium...120
Berkelium...202
Beryllium...16
Bismuth...174
Bohrium...212
Boron...18
Bromine...78
Cadmium...104
Calcium...48
Californium...203
Carbon...20
Cerium...124
Caesium...118
Chlorine...42
Chromium...56
Cobalt...62
Copernicium...217

Copper...66
Curium...200
Darmstadtium...215
Dubnium...210
Dysprosium...140
Einsteinium...204
Erbium...144
Europium...134
Fermium...205
Flerovium...219
Fluorine...26
Francium...182
Gadolinium...136
Gallium...70
Germanium...72
Gold...166
Hafnium...152
Hassium...213
Helium...12
Holmium...142
Hydrogen...10
Indium...106
Iodine...114
Iridium...162

Iron...60
Krypton...80
Lanthanum...122
Lawrencium...208
Lead...172
Lithium...14
Livermorium...221
Lutetium...150
Magnesium...32
Manganese...58
Meitnerium...214
Mendelevium...206
Mercury...168
Molybdenum...92
Moscovium...220
Neodymium...128
Neon...28
Neptunium...194
Nickel...64
Nihonium...218
Niobium...90
Nitrogen...22
Nobelium...207
Oganesson...223

Osmium...160
Oxygen...24
Palladium...100
Phosphorus...38
Platinum...164
Plutonium...196
Polonium...176
Potassium...46
Praseodymium...126
Promethium...130
Protactinium...190
Radium...184
Radon...180
Rhenium...158
Rhodium...98
Roentgenium...216
Rubidium...82
Ruthenium...96
Rutherfordium...209
Samarium...132
Scandium...50
Seaborgium...211
Selenium...76
Silicon...36

Silver...102
Sodium...30
Strontium...84
Sulfur...40
Tantalum...154
Technetium...94
Tellurium...112
Tennessine...222
Terbium...138
Thallium...170
Thorium...188
Thulium...146
Tin...108
Titanium...52
Tungsten...156
Uranium...192
Vanadium...54
Xenon...116
Ytterbium...148
Yttrium...86
Zinc...68
Zirconium...88

# Index of elements (by group)

**Actinides**
Actinium...186
Americium...198
Berkelium...202
Californium...203
Curium...200
Einsteinium...204
Fermium...205
Lawrencium...208
Mendelevium...206
Neptunium...194
Nobelium...207
Plutonium...196
Protactinium...190
Thorium...188
Uranium...192

**Alkali metals**
Caesium...118
Francium...182
Lithium...14
Potassium...46
Rubidium...82
Sodium...30

**Alkaline earth metals**
Barium...120
Beryllium...16
Calcium...48
Magnesium...32
Radium...184
Strontium...84

**Lanthanides**
Cerium...124
Dysprosium...140
Erbium...144
Europium...134
Gadolinium...136
Holmium...142
Lanthanum...122
Lutetium...150
Neodymium...128
Praseodymium...126
Promethium...130
Samarium...132
Terbium...138
Thulium...146
Ytterbium...148

**Metalloids**
Antimony...110
Arsenic...74
Boron...18
Germanium...72
Silicon...36
Tellurium...112

**Noble gases**
Argon...44
Helium...12
Krypton...80
Neon...28
Oganesson...223
Radon...180
Xenon...116

**Post-transition metals**
Aluminium...34
Astatine...178
Bismuth...174
Gallium...70
Indium...106
Lead...172
Polonium...176
Thallium...170
Tin...108

**Reactive non-metals**
Bromine...78
Carbon...20
Chlorine...42
Fluorine...26
Hydrogen...10
Iodine...114
Nitrogen...22
Oxygen...24
Phosphorus...38
Selenium...76
Sulfur...40

**Transition metals**
Bohrium...212
Cadmium...104
Chromium...56
Cobalt...62
Copper...66
Dubnium...210
Gold...166
Hafnium...152
Hassium...213
Iridium...162
Iron...60
Manganese...58
Mercury...168
Molybdenum...92
Nickel...64
Niobium...90
Osmium...160
Palladium...100
Platinum...164
Rhenium...158
Rhodium...98
Ruthenium...96
Rutherfordium...209
Scandium...50
Seaborgium...211
Silver...102
Tantalum...154
Technetium...94
Titanium...52
Tungsten...156
Vanadium...54
Yttrium...86
Zinc...68
Zirconium...88

**Unknown Properties**
Copernicium...217
Darmstadtium...215
Flerovium...219
Livermorium...221
Meitnerium...214
Moscovium...220
Nihonium...218
Roentgenium...216
Tennessine...222

# Picture credits